Green Energy and Technology

Climate change, environmental impact and the limited natural resources urge scientific research and novel technical solutions. The monograph series Green Energy and Technology serves as a publishing platform for scientific and technological approaches to "green"—i.e. environmentally friendly and sustainable—technologies. While a focus lies on energy and power supply, it also covers "green" solutions in industrial engineering and engineering design. Green Energy and Technology addresses researchers, advanced students, technical consultants as well as decision makers in industries and politics. Hence, the level of presentation spans from instructional to highly technical.

More information about this series at http://www.springer.com/series/8059

Priyadarsini Rajagopalan
Mary Myla Andamon · Trivess Moore
Editors

Energy Performance in the Australian Built Environment

Springer

Editors
Priyadarsini Rajagopalan
School of Property, Construction and Project Management
RMIT University
Melbourne, VIC
Australia

Trivess Moore
School of Property, Construction and Project Management
RMIT University
Melbourne, VIC
Australia

Mary Myla Andamon
School of Property, Construction and Project Management
RMIT University
Melbourne, VIC
Australia

ISSN 1865-3529 ISSN 1865-3537 (electronic)
Green Energy and Technology
ISBN 978-981-10-7879-8 ISBN 978-981-10-7880-4 (eBook)
https://doi.org/10.1007/978-981-10-7880-4

Library of Congress Control Number: 2018942621

Printed on acid-free paper

This Springer imprint is published by the registered company Springer Nature Singapore Pte Ltd.
The registered company address is: 152 Beach Road, #21-01/04 Gateway East, Singapore 189721, Singapore

Preface

An important function of buildings is for safety and providing healthy shelter. However, buildings as shelters have evolved and are now synonymous with providing comfort, which pose challenges inextricably linked with building performance. In a world where the supply of fossil fuels is scarcer and more expensive as well as dealing with the effects of an extreme and changing climate, energy performance of the built environment has become an essential concern.

This book stemmed from the interest of the *Sustainable Building Innovation Laboratory* (SBi Lab) research group to research and report on how the Australian built environment is placed within the wider context of building energy performance. The book is intended to take readers on a tour of policy, technology, design and case studies of energy performance in built environment developments in Australia. It covers a range of built environment types including residential and non-residential buildings.

The aim of this book is to promote understanding and incite discussion of environmentally sustainable approaches to building development in Australia with a view to a lower carbon society. Some chapters are technical, while others are more general in nature. In doing this, the book is relevant to policy makers and researchers in this space, but it is also designed to introduce readers who may be interested in a sustainable built environment to key developments, challenges, case studies and opportunities. The chapters in this book contribute to a growing Australian and international literature on how to develop a more energy efficient, and ultimately more sustainable, built environment.

Melbourne, Australia

Priyadarsini Rajagopalan
Mary Myla Andamon
Trivess Moore

Contents

Editors and Contributors

About the Editors

Assoc. Prof. Priyadarsini Rajagopalan, B.Tech. Arch. Eng., MBldSc, Ph.D. is an associate professor with expertise in building energy efficiency and energy benchmarking, indoor environmental performance and urban microclimate.

Dr. Mary Myla Andamon, B.Sc. Arch., Ph.D. is a lecturer and a building scientist with expertise in indoor and outdoor thermal comfort studies, urban microclimate and building environmental monitoring and performance assessments.

Dr. Trivess Moore, B.Sc. (Env.), Ph.D. is a research fellow with expertise in energy efficiency, renewable energy, the built environment and transitioning to a low-carbon future.

Contributors

Mary Myla Andamon Sustainable Building Innovation Laboratory, School of Property Construction and Project Management, RMIT University, Melbourne, VIC, Australia

Andrew Carre Sustainable Building Innovation Laboratory, School of Property Construction and Project Management, RMIT University, Melbourne, VIC, Australia

Ruwini Edirisinghe Sustainable Building Innovation Laboratory, School of Property Construction and Project Management, RMIT University, Melbourne, VIC, Australia

Matthew Francis Sustainable Building Innovation Laboratory, School of Property Construction and Project Management, RMIT University, Melbourne, VIC, Australia

Sarah Holdsworth Sustainable Building Innovation Laboratory, School of Property Construction and Project Management, RMIT University, Melbourne, VIC, Australia

Trivess Moore Sustainable Building Innovation Laboratory, School of Property Construction and Project Management, RMIT University, Melbourne, VIC, Australia

Priyadarsini Rajagopalan Sustainable Building Innovation Laboratory, School of Property Construction and Project Management, RMIT University, Melbourne, VIC, Australia

Jin Woo Sustainable Building Innovation Laboratory, School of Property Construction and Project Management, RMIT University, Melbourne, VIC, Australia

Rebecca J. Yang Sustainable Building Innovation Laboratory, School of Property Construction and Project Management, RMIT University, Melbourne, VIC, Australia

Affiliates

Adisa Adejare Alawode School of Property Construction and Project Management, RMIT University, Melbourne, VIC, Australia

Jeremy Cooke School of Property Construction and Project Management, RMIT University, Melbourne, VIC, Australia

David Kenny School of Property Construction and Project Management, RMIT University, Melbourne, VIC, Australia

Shaun Matfin School of Property Construction and Project Management, RMIT University, Melbourne, VIC, Australia

The Built Environment in Australia

Priyadarsini Rajagopalan, Trivess Moore and Mary Myla Andamon

The pressures from climate change, population growth and other social, health, well-being, livability, usability and affordability factors on Australia's built environment are significant and complex, as it is the with many developed and developing countries around the world [24]. Increasing evidence from around the world is demonstrating that improving the environmental sustainability of our built environment can help to address a number of these elements such as reducing environmental impacts, improving occupant health and reducing operating costs [5, 18, 26–28]. International best practice is now for buildings that have low- or net zero-carbon (or environmental) impact [19], but there are also examples of buildings which intentionally go beyond this to help offset the impacts caused by other buildings, as demonstrated through the Living Building Challenge framework (see https://living-future.org/).

This chapter outlines the state of play of energy performance of the built environment in Australia and places it within the global context. Despite many examples of improved buildings and outcomes for the environment, occupants and society, most new and existing buildings around the world fall significantly short of such low-/zero-carbon performance outcomes [23]. This is cause for concern as we transition towards a low-carbon future, with the globally scientific and political consensus that we must take urgent action to reduce our greenhouse gas emissions if we are to mitigate significant climate change outcomes. The built environment remains a significant contributor of greenhouse gas emissions (19% globally) and accounts for 32% of final energy demand globally [17]. The good news is that the built environment has been identified as having substantial "low hanging fruits"—primarily in the way of improving energy performance, much of which have been identified as being cost efficient [14, 17, 21]. Major reports from around the world continue to emerge which demonstrate that cities, regions or even countries can transition to a low- or zero-carbon built environment by 2050 if not sooner [6, 8, 19]; however, this will take significant innovation and coordinated action from key government and industry stakeholders if it is to be achieved.

P. Rajagopalan (✉) · T. Moore · M. M. Andamon
Sustainable Building Innovation Laboratory, School of Property Construction and Project Management, RMIT University, Melbourne, VIC, Australia
e-mail: priyadarsini.rajagopalan@rmit.edu.au

P. Rajagopalan et al. (eds.), *Energy Performance in the Australian Built Environment*, Green Energy and Technology, https://doi.org/10.1007/978-981-10-7880-4_1

While such research shows what can be achieved, the reality of implementing such outcomes remains contested due to a number of factors. Firstly, there has been a shift over the past 30 years or so towards regulating minimum performance standards in the built environment [4, 9, 15, 22, 25]. This emerged globally as a way to lift the bottom quality of the market and ensure a base-level performance across all buildings. Over time, the minimum requirements have been strengthened and while they have undoubtedly improved the performance of the built environment, particularly through improvements to energy efficiency [2, 20], they still fall short of what is required for a low-carbon future. In countries like Australia, the majority of the built environment is delivered to meet minimum standards and often not even achieving that.

Secondly, the building industry has historically been a staunch critic of anything they perceive to add additional costs to their product [10]. Many building industry stakeholders have argued that if there are additional requirements added to minimum performance regulations, that will impact on capital costs and ultimately be passed onto consumers [13]. In a time where property prices have exploded over the past 15 years or so, affordability of the built environment, and housing in particular, has become an important political issue, not only in Australia but globally.

Thirdly, these key stakeholders argue that the market will decide if there is a value or place for improved sustainability. This neoclassical economics views three key assumptions about consumers: that consumers use rationality in their decision development, that consumers make decisions which maximise the outcome for themselves and that consumers make these decisions independently, based on complete information. Clearly as the limited action on climate change has shown, there is a significant market failure when it comes to consumers and taking action. This is evident in the built environmental sector as well, with only limited numbers of buildings approaching the performance required for a low-carbon future.

This book provides a discussion of the state of play of energy performance in the Australian built environment. Australia is similar in many ways to other OECD countries in relation to governance, economic development and the challenges faced in transitioning the built environment to a low-carbon future. Australia has become highly urbanised and faces a rapid population increase over the coming decades with a predicted population increase of almost 50% (from 2016 level) over the next three decades [16]. This increasing population highlights the challenge of providing a more sustainable built environment while trying to maintain certain quality of life outcomes. Although, there are many who argue a sustainable future will require degrowth [1]. While this book is firmly focused on research and policy development in Australia, it does attempt to locate this into a relevant international discussion.

By way of context, Australia's built environment contributes 26% of greenhouse gas emissions and consumes 19% of total energy [3]. The combined residential and commercial sector accounted for 17% of final energy consumption in 2014–15 [7]. Between 1990 and 2020, the number of occupied residential households will increase from six million to almost 10 million, an increase of 61% [12]. Total energy consumption in residential and commercial buildings is expected to rise to 467 PJ and 170 PJ, respectively, by 2020 [11]. Among non-residential buildings, schools show the lowest energy use per square metre on average (176 MJ/m^2) whereas hotels, hospitals and shopping centres show above 1000 MJ/m^2, and supermarkets showed the highest energy intensity at over 3300 MJ/m^2 [11]. Heating, ventilation and air conditioning (HVAC) is generally the largest end use of electricity, with lighting and equipment following behind, while space heating is the dominant end use for gas. However, this changes across the country with different climate zones impacting on energy consumption.

This book is structured in four main parts. Part I, *Climate and Impacts,* discusses the key challenges relating to the climate, the changing urban climate due to climate change and outdoor thermal environments. Chapter "Urban Climates in the Transformation of Australian Cities" attempts to understand urban climates in cities to manipulate spaces and buildings and create better city environments. Chapter "Thermal Environments in the Construction Industry: A Critical Review of Heat Stress Assessment and Control Strategies" builds upon this by discussing the main problems and risks associated with heat stress (an outcome of changing city microclimates and thermal environments), with an emphasis on the construction industry. Part II, *Regulatory Frameworks*, includes two chapters reviewing the development and status of approaches to improve energy efficiency and broader sustainability in the Australian built environment. Chapter "The Built Environment and Energy Efficiency in Australia: Current State of Play and Where to Next" discusses the minimum building performance requirements set through the National Construction Code–Building Code of Australia as well as other mandatory and voluntary approaches which have been introduced over the past two decades to improve energy efficiency and broader sustainability in new and renovated buildings. Chapter "Environmental Rating Systems for Non-Residential Buildings—How Does Australia Compare with International Best Practice?" reviews the status of environmental rating systems in the non-residential building sector in Australia and compares this with other leading international rating systems. Part III, *Case Studies*, then presents a series of different case studies, respectively, covering multi-residential development (Chapter "An End-User-Focused Building Energy Audit: A High-Density Multi-residential Development in Melbourne, Australia"), detached social housing (Chapter "Low-Energy Housing as a Means of Improved Social Housing: Benefits, Challenges and Opportunities"), primary and secondary school educational facilities (Chapter "Indoor Environmental Quality of Preparatory to Year 12 (P-12) Educational Facilities in Australia: Challenges and Prospects"), university buildings (Chapter "University Buildings: The Push and Pull for Sustainability"), indoor aquatic centres (Chapter "A Guide for Evaluating the Performance of Indoor Aquatic Centres") and distributed solar systems (Chapter " A Feasibility Study and Assessment: Distributed Solar System in High-Density Areas"). Part IV, *Future Direction and Imperatives*, includes three chapters to round out the book. Chapter "Are We Living with Our Heads in the Clouds? Perceptions of Liveability in the Melbourne High-Rise Apartment Market" examines the characteristics of liveability and design in the context of high-rise residential developments highlighting key considerations for future developments. Chapter "The Way Forward—Moving Toward Net Zero Energy Standards" explores net zero energy standards and discusses how Australia is positioned among comparative countries and the ways to move forward towards this low-carbon future. Chapter "Cohesion: Our Environment—Building Better and Smarter" then provides a short narrative of the book's key themes and puts forward arguments for policy-makers, building industry stakeholders and researchers to do more to drive the sustainable built environmental agenda in Australia and indeed globally.

1 Part I: Climate and Impact

Part I presents three chapters which look at different aspects of climate change and the impacts on urban climate, outdoor thermal environments and building carbon footprint. Chapter "Urban Climates in the Transformation of Australian Cities" by Andamon and Carre presents a review of the changing climate in Australia and how this is transforming Australian cities, its urban spaces and built environments. The authors contend that understanding the urban climate and its interaction with the built environment is key to the successful application of strategies to mitigate the adverse impacts of human-caused climate change. In Chapter "Thermal Environments in the Construction Industry: A Critical Review of Heat Stress Assessment and Control Strategies", Edirisinghe and Andamon build on the discussion on the impacts of extreme weather conditions brought by climate change and explore this within the context of thermal environments in construction industry settings. The authors present a discussion on the occupational health hazards for the construction industry workers at risk of extreme heat exposure and the strategies and controls to mitigate the impact of heat stress.

2 Part II: Regulatory Frameworks

Chapter "The Built Environment and Energy Efficiency in Australia: Current State of Play and Where to Next" by Moore and Holdsworth presents a review of approaches to improve energy efficiency and broader environmental sustainability in the Australian built environment. The authors discuss the minimum building performance requirements set through the National Construction Code, but also explore other mandatory and voluntary approaches which have been introduced over the past two decades. The chapter concludes with a discussion that highlights current gaps relating to the delivery of a low-carbon built environment. The authors argue that Australia currently fails to meet international building performance best practice standards and this situation can only be reversed if various levels of government in Australia increase the regulated level of energy performance of buildings coupled with a more holistic and progressive inclusion in rating tools for all energy consumed and generated within a building.

In Chapter "Environmental Rating Systems for Non-Residential Buildings—How Does Australia Compare with International Best Practice?", Rajagopalan builds upon Chapter "The Built Environment and Energy Efficiency in Australia: Current State of Play and Where to Next" but providing a more detailed look at what is happening in the non-residential sector in Australia and internationally. The chapter talks about the impact that the largest non-residential rating system in Australia, Green Star, has had on the building industry. This discussion about performance and market penetration is compared to what is happening with LEED and BREEAM. Rajagopalan finds that even though Green Star has similar criteria and performance standards in comparison with LEED and BREEAM, the market penetration of this rating system falls behind other systems in terms of adoption rate. She argues that proper government support and improvement of supply chains would certainly help the rating systems to penetrate the wider market.

3 Part III: Case Studies

This part of the book explores six different case studies of improving energy efficiency and performance in the Australian built environment. In Chapter "An End-User-Focused Building Energy Audit: A High-Density Multi-residential Development in Melbourne, Australia", Woo and Moore demonstrate a building energy audit process using a case study of high-density multi-residential modular development in inner Melbourne, Australia. The authors argue that an energy audit is essential to understand where and how energy is used in buildings and consequently to identify those areas where improvements can be made but that there are limited studies of higher density housing in Australia which draws upon real energy consumption data and occupant feedback. They found that the occupants raised issues with poor thermal discomfort in summer. Energy consumption was found to be significantly less than the average consumption in the same suburb and more likely to be affected by housing tenure types than physical building conditions such as orientation and height. In Chapter "Low-Energy Housing as a Means of Improved Social Housing: Benefits, Challenges and Opportunities", Moore presents outcomes of a multi-year evaluation of a cohort of low-energy social housing from Horsham in regional Victoria, Australia. The analysis includes technical performance data and is supplemented with the occupants' own stories about improved livability outcomes. The evidence finds that social housing providers should consider providing homes which go beyond minimum building regulations as they provide a range of benefits for energy consumption as well as improving occupant health and well-being. In addition, Moore argues that such housing would help social housing providers achieve organisational or broader government sustainability goals such as reducing greenhouse gas emissions and fossil fuel energy consumption.

Chapter "Indoor Environmental Quality of Preparatory to Year 12 (P-12) Educational Facilities in Australia: Challenges and Prospects" by Andamon and Woo focuses on primary and secondary educational facilities and reviews the indoor environmental performance of school buildings by looking at national and international standards, design guidelines and policies on indoor environmental quality (IEQ). School indoor environments are particularly at risk of the impact of extreme weather conditions. The authors argue that studies on indoor environmental conditions of Australian schools are needed, backed by measurements and surveys of comfort conditions and the relationship between quality of indoor environments and student performance.

This is followed by Chapter "University Buildings: The Push and Pull for Sustainability" where Francis and Moore explore the role that universities in Australia are playing in a transition to a low-carbon, energy efficient future. After discussing relevant policies and rating tools, the authors present five key examples that go significantly beyond minimum performance requirements from prominent Australian universities. Evident from the examples is that there continues to be no one-size-fits-all approach for universities to become more sustainable.

In Chapter "A Guide for Evaluating the Performance of Indoor Aquatic Centres", Rajagopalan explores the case of aquatic centres which are complex building types accommodating diverse facilities that are distinct in their functional requirements. Rajagopalan provides an overview of the characteristics of aquatic centres, highlighting the challenges in evaluating the performance of these buildings, and proposes a methodology for evaluating the design and operational performance of these buildings. Completing this section is Chapter "A Feasibility Study and Assessment: Distributed Solar System in High-Density Areas" by Yang and Carre. They present a case study which details a value assessment to optimise the cost of applying solar photovoltaic systems in a high-density city area of Melbourne to help better inform investment decisions.

4 Part IV: Future Directions and Imperatives

In Chapter "Are We Living with Our Heads in the Clouds? Perceptions of Liveability in the Melbourne High-Rise Apartment Market", Holdsworth et al. examine the characteristics of liveability and design in the context of high-rise residential developments in Melbourne, Australia. The chapter includes considerations of building amenity, apartment amenity and external amenity. Through occupant interviews, the authors explore perceptions of liveability as they inform and consequently manifest in current development projects. The findings identified that liveability is a subjective term encompassing a variety of characteristics which different stakeholder groups emphasised differently based on their disciplinary background. The authors argue that the findings are important as there exists limited understanding of how the building industry conceptualises high-rise developments and in turn makes design and development decisions in the context of liveability. Further, it was recognised that all participants wanted to improve the liveability of their development and were prepared to collaborate across discipline to achieve such outcomes.

Following this, Chapter "The Way Forward—Moving Toward Net Zero Energy Standards" reviews recent advances in the high-performance building standards with emphasis on global developments of net zero energy standards, discusses how Australia is positioned in relation to this standard and found that Australia is yet to formulate a policy towards adopting a net zero energy building standards. Many scholars recognise that Australia cannot delay the implementation of deep improvements in energy efficiency in the built environment any longer, as issues of energy security, affordability and increasing greenhouse gas emissions have become critical.

A short conclusion (Chapter "Cohesion: Our Environment—Building Better and Smarter") then brings the book to a close by highlighting key themes from the book and discussing them within the urgent need to improve the energy performance of our built environment, both in Australia and globally, through a more collaborative approach.

References

1. Alexander S, Rutherford J (2018) The 'transition town' movement as a model for urban transformation. In: Moore T, de Haan F, Horne R, Gleeson B (eds) Urban sustainability transitions. Theory and practice of urban sustainability transitions. Springer, Singapore
2. Ambrose M, James M, Law A, Osman P, White S (2013) The evaluation of the 5-star energy efficiency standard for residential buildings. Retrieved from Sydney
3. ASBEC (2013) The second plank—building a low carbon economy with energy efficient buildings: http://www.asbec.asn.au/files/ASBEC%20CCTG%20Second%20Plank%20Report%202.0_0.pdf
4. Berry S, Marker T (2015) Residential energy efficiency standards in Australia: where to next? Energ Effi 8(5):963–974. https://doi.org/10.1007/s12053-015-9336-4
5. Berry S, Whaley D, Davidson K, Saman W (2014) Do the numbers stack up? Lessons from a zero carbon housing estate. Renew Energy 67:80–89. https://doi.org/10.1016/j.renene.2013.11.031
6. Boardman B, Darby S, Killip G, Hinnells M, Jardine N, Palmer J, Sinden G (2005) 40% house: Environmental Change Institute, University of Oxford
7. BREE (2014) Australian Energy Projections to 2049-50, Commonwealth of Australia: https://industry.gov.au/Office-of-the-Chief-Economist/Publications/Documents/aep/aep-2014-v2.pdf
8. BZE (2013) Zero carbon Australia: buildings plan. Retrieved from Melbourne: http://bze.org.au/buildings-plan/
9. CBSC (2010) 2010 California green building standards code. California code of regulations, Title 24, Part 11. USA: California Building Standards Commission
10. Crabtree L, Hes D (2009) Sustainability uptake on housing in metropolitan Australia: an institutional problem, not a technological one. Housing Stud 24(2):203–224. https://doi.org/10.1080/02673030802704337
11. DCCEE (2012) Baseline energy consumption and greenhouse gas emissions in commercial buildings in Australia. Retrieved from Canberra: https://industry.gov.au/Energy/EnergyEfficiency/Non-residentialBuildings/Documents/CBBS-Part-1.pdf
12. DEWHA (2008) Energy use in the Australian residential sector 1986–2020. Department of the Environment Water Heritage and the Arts, Canberra
13. HIA (2009) Response to the consultation regulation impact statement 2009–03. Proposal to revise the energy efficiency requirements of the BCA for residential buildings. Retrieved from Campbell, ACT

14. Hoppe T, Bressers J, Lulofs K (2011) Local government influence on energy conservation ambitions in existing housing sites—plucking the low-hanging fruit? Energy Policy 39(2):916–925. https://doi.org/10.1016/j.enpol.2010.11.016
15. Horne R, Hayles C (2008) Towards global benchmarking for sustainable homes: an international comparison of the energy performance of housing. J Housing Built Environ 23(2):119–130. https://doi.org/10.1007/s10901-008-9105-1
16. Infrastructure Australia, n.p. (2015) Population estimates and projections. Australian Infrastructure audit background paper April 2015. Retrieved from Canberra: http://infrastructureaustralia.gov.au/policy-publications/publications/files/Background-paper-on-demographic-projections.pdf
17. IPCC (2014) Fifth Assessment Report. Climate Change 2014: Working Group III: Mitigation of Climate Change. Retrieved from Valencia, Spain
18. Kats G (2009) Greening our built world: costs, benefits, and strategies. Island Press, Washington, DC
19. Kolokotsa D, Rovas D, Kosmatopoulos E, Kalaitzakis K (2011) A roadmap towards intelligent net zero- and positive-energy buildings. Sol Energy 85(12):3067–3084. https://doi.org/10.1016/j.solener.2010.09.001
20. Levinson A (2014) How much energy do building energy codes really save? Evidence from California. NBER Working Paper No. w20797. Retrieved from Washington
21. Lovell H (2004) Framing sustainable housing as a solution to climate change. J Environ Planning Policy Manage 6(1):35–55. https://doi.org/10.1080/1523908042000259677
22. Lowe R, Oreszczyn T (2008) Regulatory standards and barriers to improved performance for housing. Energy Policy 36(12):4475–4481. https://doi.org/10.1016/j.enpol.2008.09.024
23. Moore T (2012) Facilitating a transition to zero emission new housing in Australia: costs, benefits and direction for policy. RMIT University, Melbourne
24. Moore T, de Haan F, Horne R, Gleeson B (2018) Urban sustainability transitions. Australian cases—international perspectives. Springer, Singapore
25. Moore T, Morrissey J (2014) Lifecycle costing sensitivities for zero energy housing in Melbourne, Australia. Energy Build 79:1–11. https://doi.org/10.1016/j.enbuild.2014.04.050
26. Newton P (2012) Liveable and sustainable? Socio-technical challenges for twenty-first-century cities. J Urban Technol 19(1):81–102. https://doi.org/10.1080/10630732.2012.626703
27. Willand N, Ridley I, Maller C (2015) Towards explaining the health impacts of residential energy efficiency interventions—a realist review. Part 1: Pathways. Soc Sci Med 133(Supplement C):191–201. https://doi.org/10.1016/j.socscimed.2015.02.005
28. Zabaneh G (2011) Zero net house: preliminary assessment of suitability for Alberta. Renew Sustain Energy Rev 15(6):3237–3242. https://doi.org/10.1016/j.rser.2011.04.021

Assoc. Prof. Priyadarsini Rajagopalan, B.Tech. Arch. Eng., M.Bld.Sc., Ph.D. is an associate professor with expertise in building energy efficiency and energy benchmarking, indoor environmental performance and urban microclimate.

Dr. Trivess Moore, B.Sc. (Env), Ph.D. is a research fellow with expertise in energy efficiency, renewable energies, the built environment and transitioning to a low carbon future.

Dr. Mary Myla Andamon, B.Sc. Arch., Ph.D. is a lecturer and a building scientist with expertise in indoor and outdoor thermal comfort studies, urban microclimate and building environmental monitoring and performance assessments.

Urban Climates in the Transformation of Australian Cities

Mary Myla Andamon and Andrew Carre

Abstract Climate change is set to significantly impact cities and those who live and work within them. This chapter reviews the changing climate in Australia and its consequent role in the transformation of Australian cities with emphasis on the impact to the built environment. Following this, a discussion explores the application of strategies to mitigate the adverse impacts of climate change on buildings and cities. While these strategies will be important for a transition to a low-carbon future, there is still a requirement for innovative research and developments that would pave clearer directions to achieve a lower carbon society.

1 Australia's Changing Climate

Australia is the world's second-driest continent (after Antarctica) and features a wide range of climatic zones, from the tropical regions of the north, through the arid expanses of the interior, to the temperate regions of the south. The climate in this island continent is largely determined by its latitude lying between 10°S and 39°S. Except for the state of Tasmania, Australia's low and flat terrain and generally low relief mean that topography has less impact on atmospheric systems that control the climate compared to more mountainous continents [1]. Australia experiences many of nature's more extreme weather phenomena, including droughts, floods, tropical cyclones, severe storms and bush fires.

Australia's temperatures were relatively stable since national records began in 1910–1950 [1]. From 1950, however, records have shown an increasing trend in both minimum and maximum temperatures where very warm months that occurred just over 2% of the time during the period 1951–1980, occurred nearly 7% of the time during 1981–2010, and around 10% of the time over the past 15 years [18]. At the same time, the frequency of very cool months has declined by around a third

M. M. Andamon (✉) · A. Carre
Sustainable Building Innovation Laboratory, School of Property Construction and Project Management, RMIT University, Melbourne, VIC, Australia
e-mail: mary.andamon@rmit.edu.au

P. Rajagopalan et al. (eds.), *Energy Performance in the Australian Built Environment*, Green Energy and Technology, https://doi.org/10.1007/978-981-10-7880-4_2

since the earlier period. Mean surface air temperatures have increased by 0.9 °C since 1910. Daytime maximum temperatures have increased by 0.8 °C over the same period, while overnight minimum temperatures have warmed by 1.1 °C [19].

1.1 The Global Picture: Warming and Extreme Weather Events

Although natural variability plays an important role in the occurrence of weather and climate extremes [44], climate scientists agree that human influence on climate from increasing greenhouse gases into the atmosphere is a main contributor to global warming. Extreme event attribution is a relatively new field gaining momentum in response to public interest in linking the seemingly abstract concept of human-caused climate change or natural climatic variability with tangible experiences of the damaging extreme weather events [48]. In 138 attribution studies conducted on 144 extreme weather events over the past 20 years reviewed and mapped by Carbon Brief, 68% of the cases have shown that "*human-caused climate change has altered the likelihood or severity of the extreme weather event*" [57, p. 3]).

The literature analysed in the Carbon Brief mapping is dominated by studies of extreme heat (34%), droughts (23%) and heavy rainfall or floods (20%) (Fig. 1). Hurricane Sandy along the eastern US seaboard in 2012 and Typhoon Haiyan in the Philippines in 2013 were made more severe or likely to occur because of human influence. Anthropogenic climate change-induced rainfall trends were also found to be a significant contributor to the most severe Australian decade-long "Millennium Drought" in 2001–2010/2011 [10], the California drought in 2013–2014 [78] and drought in Tasmania, Australia, in 2015 [38]. Extreme rainfall in India in June 2013 [71] and Northland, New Zealand, in July 2014 [62] was also attributed to anthropogenic influence on climate.

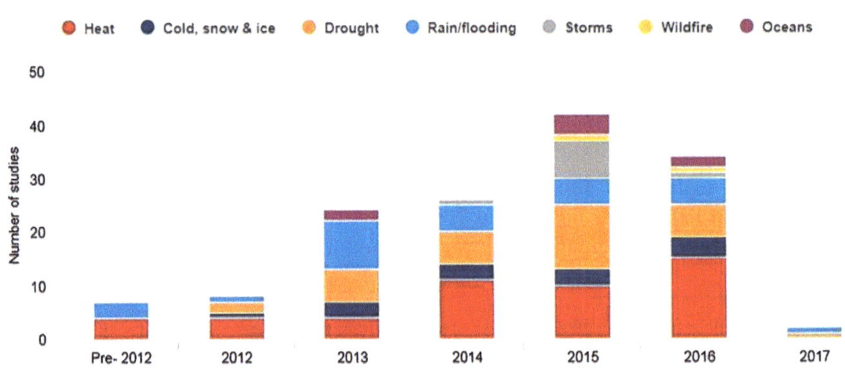

Fig. 1 Extreme weather event attribution studies (*Source* Carbon Brief [57, p. 4])

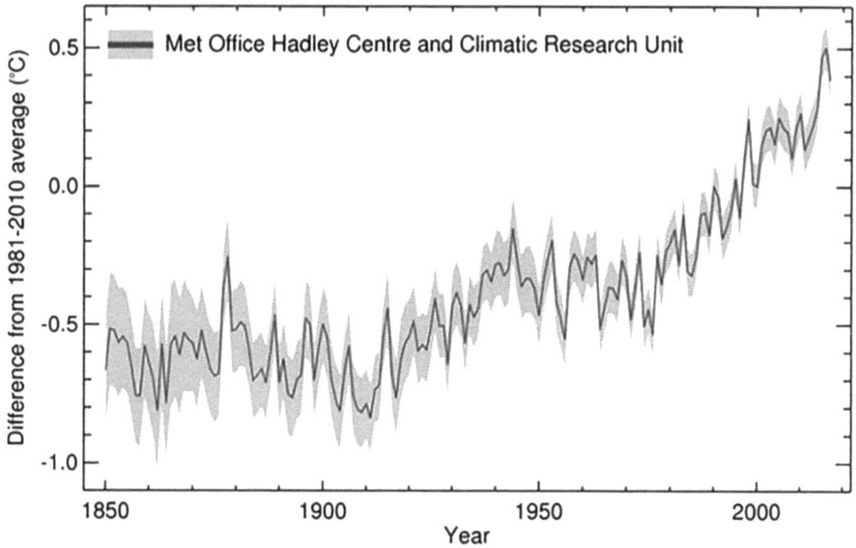

Fig. 2 Global average temperature anomaly, 1850–2017 (*Source* UK Met Office Hadley Centre and Climatic Research Unit [77])

1.2 Rising and Extreme Heat is Defining Climate Change

Globally, 2015–2017 were the three warmest years on record [81]. Global average temperatures in 2017 and 2015 were 1.1 °C above pre-industrial levels (Fig. 2). While 2016 is the warmest year on record with average surface temperatures 1.2 °C above the pre-industrial years, the globally averaged temperature of 14.7 °C for 2017 makes it the warmest without the influence of warming from El Niño [77]. The El Niño event which spanned 2015–2016 contributed to around 0.2 °C to the annual average in 2016.

The Carbon Brief review of 48 attribution studies which looked at extreme heat around the world found that 85% of the heatwave studies have been altered by climate change [57]. With 25% of the heat-related attribution studies, Australia is the most studied region for heatwaves in the Carbon Brief review demonstrating the risks that Australia continues to face with a changing climate.

1.3 Australian Angry Summers, Abnormal Autumns, Warm–Dry Winters and Scorching Springs

Although Australia has always had heatwaves, hot days and bush fires, climate change is increasing the risk of more frequent and longer heatwaves and more extreme hot days [13]. This was demonstrated by the extreme weather events which dominated

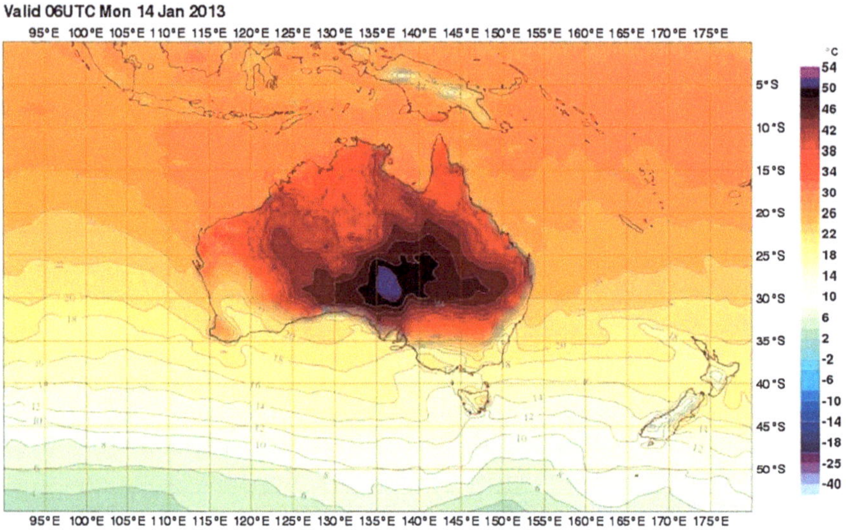

Fig. 3 BoM temperature forecast for 14 January 2013 [31] (*Source* Australian Bureau of Meteorology)

the 2012/2013 Australian summer. Weather records were broken over a 90-day period include the hottest day ever recorded for Australia as a whole: the hottest January on record, the hottest summer average on record, and a record seven days in a row when the whole continent averaged above 39 °C [6]. 2012/2013 summer has then been referred to as the *Extreme Summer* or *Angry Summer* [12]. The extreme maximum temperatures set a new national average maximum of 40.33 °C on 7 January 2013 which prompted the Australian Bureau of Meteorology (BoM) to alter the weather forecasting chart [33]. The previous temperature range that had been capped at 50 °C scale now extends to 54 °C with the addition of new colours deep purple (52 °C) and pink (54 °C) (Fig. 3).

Climate change is making Australian cities and major regions hotter [8]. It would seem that angry summers have become the norm. Hot temperatures are experienced more often, and heatwaves are becoming hotter, longer and more frequent [73]. Across Australia, these rising and extreme summer temperatures have extended to the other seasons as well. In 2017, the autumn mean maximum temperatures were above average (+1.21 °C), winter was the warmest on record with average maximum temperatures nearly 2 °C above average, and Melbourne had 15 days of 30 °C or more during spring [74].

2 The Australian Story: Impacts of Extreme Heat and Warm, Drier Climate

Australia is a highly urbanised country. Although much of the major cities are located along the coastline, the populations across all the major cities are at record levels [25]. The major human impacts of heatwaves are on human health, productivity, energy and infrastructure [21, 29, 34]. Extreme heat is a crucial cause of morbidity and mortality. Major heatwaves have now become Australia's deadliest natural hazards having caused more deaths than bush fires, cyclones and earthquakes since 1890 [14]. Often called the "silent killer", extreme heat is a serious health threat for many Australians where more die every year from extreme heat events (EHE) than from any other type of natural disaster [58]. People living in the dense urban environments are particularly at risk from the effects of heatwaves when temperatures soar [32]. For example, the 2009 heatwave in south-east Australia caused over 400 deaths in Melbourne and Adelaide alone [26].

The examination of four decades (1968–2007) of Australia's mortality data indicated that there has been a steady increase in the number of deaths in summer compared to those in winter [5]. A 10% increase in both deaths and ambulance call-outs in New South Wales from 2005 to 2015 was due to extreme heatwaves [36]. Longer, hotter and more intense heatwaves increase the risk of heat-related illness and can also exacerbate pre-existing conditions—children and the elderly are especially vulnerable [79]. Often neglected are those outdoor workers, and those working in enclosed spaces without adequate ventilation are most at risk under extreme conditions [42]. These are the construction workers, farmers, emergency and essential services workers and those working in the mining industry [63, 82]. The case of heat stress in the Australian construction industry is reviewed in Chapter "Environmental Rating Systems for Non-Residential Buildings—How Does Australia Compare with International Best Practice?".

Extreme heat is damaging to infrastructure such as electricity and transport systems [45, 49, 59, 73]. The most severe heatwave in the summer of 2017 was in south-east Australia, where daytime temperatures were above 40 °C. The highest temperature in South Australia recorded on 8 February 2017 was at Moomba where the daytime maximum reached 46.6 °C at Moomba Airport, while Adelaide reached a high of 42.4 °C [7]. This heatwave in South Australia left 40,000 people without power for about half an hour in the early evening of 8 February, while temperatures were over 40 °C [74].

Despite energy supply being available, soaring and extreme temperatures put energy systems under pressure. New South Wales also experienced the same heatwave in February 2017 where temperatures at Sydney Airport reach 42.9 °C on 10 February [7, 9]. The Australian Energy Market Operator reported that despite near record all-time peak electricity demand, NSW avoided widespread blackouts by importing electricity via interconnections with Victoria and Queensland which ran above design limits, contributing 12% to meeting peak demand, and careful use by consumers which saved 200 MW [2]. Heatwaves highlight the vulnerability of Australia's energy systems to extreme weather.

The prospect of future hot weather in Australia to happen more frequently and for longer periods due to climate change demonstrates the importance of greenhouse gas emission reduction [23] and places an urgent imperative for leadership in addressing climate change with coherent policies [74].

3 Transformation of Built Environments and Urban Spaces in Australian Cities

This chapter focuses on key effects of climate change on the built environment and urban spaces: urban heat island, consequence on building performance and energy use, damage to urban vegetation and sea level rise. People living in cities are particularly at risk of the effects of heatwaves because metropolitan areas tend to be significantly warmer than neighbouring countryside areas [16]. Urban vegetation which would have ameliorated elevated temperatures in urban areas is also affected and damaged by climate change [39]. Dense urban areas and inner-city environments may be 1–3 °C warmer than surrounding areas (Fig. 4). This phenomenon is the urban heat island (UHI), one of the most important manifestations of urban climate, where urban areas become warmer, often by several degrees than surrounding rural countryside [52]. Low-density, sprawling patterns of urban development have been also associated with enhanced surface temperatures in urbanised areas [76].

Urban heat island is generally more prominent during the night driven by heat that is trapped and stored in the urban landscape during the day and then slowly released at night [76]. A detrimental effect of UHI in many cities is the elevated temperatures which can be dangerous for some vulnerable city dwellers [30, 53, 67]. Heat stress associated with elevated temperatures has been linked to higher rates of human mortality and illness [40, 55], particularly among vulnerable demographics such as the elderly; children; lower socio-economic classes; and residents in high density, older housing stock with limited surrounding vegetation [15, 16, 60, 61]. Australians are increasingly moving into urban areas and inner cities where urban heat island effect is more likely [25].

A consequence of the increased and elevated heat loads in urban areas is the demand for space cooling in homes and buildings [70]. Cooling demands are likely to increase considerably due to rapidly expanding urban areas and extreme heat events due to climate change [46, 66]. Salamanca et al [64] suggest that cooling demand due to air-conditioning (AC) systems can consume more than 50% of the total electricity demand during extreme heat events in semi-arid urban environments, with maximum consumption up to 65% of total electricity demand during peak late afternoon hours. This is in agreement with the reported significant increase of air-conditioner use in the Australian residential sector which has effectively negated energy consumption reductions gained by improved efficiency [27]. Ownership of air-conditioning

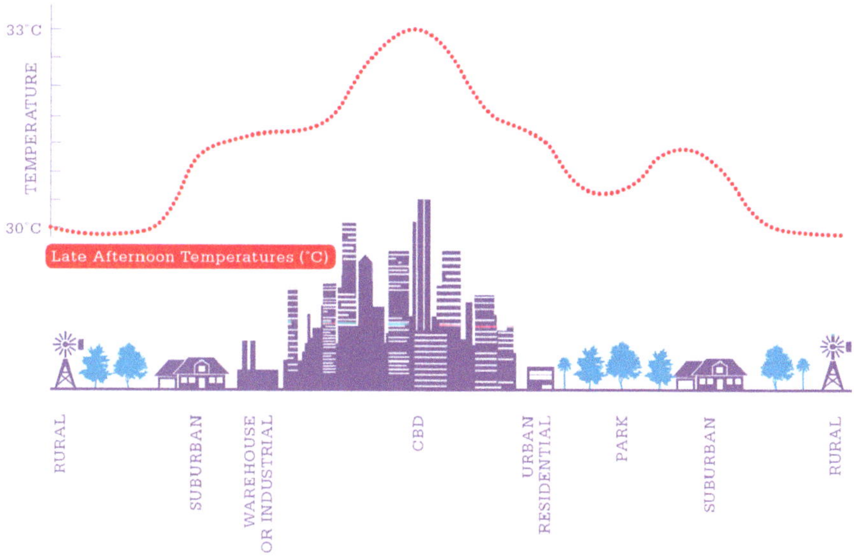

Fig. 4 Urban heat island effect. The average annual air temperature in cities (of more than one million people) may be 1–3 °C warmer than surrounding areas (*Source* Climate Commission 2011, Fig. 4, p. 12 [34])

units almost doubled between 1994 and 2004, rising from an average of 0.395 units to 0.762 units per household. The energy requirement of air-conditioning use is a major challenge particularly with climate change creating higher heat exposure levels warranting the need for more air conditioning in highly populated urban areas [47, 69].

Rising sea levels is also major consequence of climate change which affects a significant proportion of the Australian population. About 85% of Australians live within 50 kilometres off the coast [26]. More than 700,000 dwellings are within three kilometres of the coast and less than six metres above existing sea level. Projected impacts of climate change show that a significant number of residential buildings may be at risk of inundation and damage from a sea level rise of 110 cm (high-end scenario for 2100) [20]. Projections also show an increased frequency of extreme weather events with associated storm surges and coastal erosion [11, 28]. A CSIRO National Research Flagships report on climate adaptation suggests that a one-in-100-year storm tide height in Geelong is likely to rise from 110 cm to 220 cm by the end of the century [50]. The combined effect of sea level rise and a storm surge from a one-in-100-year storm would put at risk estimated 27,600–44,600 residential buildings in Victoria [20].

4 The Case for Low-Carbon (or Carbon Neutral) Built Environment

The projections for extreme weather events and particularly extreme hot days in Australia are consistent with projected global trends [73]. For both northern and southern Australia, one-in-20-year extreme hot days are expected to occur every two to five years by the middle of the century [35]. The record-breaking extreme temperatures in the summer of 2016/2017 [51, 75] and early 2018 [56] have clearly shown that the effects of human-caused climate change have become commonplace, occurring almost every summer across the country since the previous angry summer of 2012/2013 and 2013/2014 [41, 75].

Planning strategies for urban development often consider housing, transport, water and infrastructure. Yet, a very few strategies comprehensively target urban climate and its interaction with the building environment. The commonplace occurrence of extreme hot temperatures and consequent UHI effect in urban areas place an importance of "*incorporating urban climate understanding and knowledge into urban planning processes to better develop cities that are more sustainable*" and climate-sensitive, [17, p 27]. For example, the typical and prevalent solution of air conditioning (AC) poses a sustainability challenge in that increased AC use puts a burden on the electricity distribution during extreme heat days as well as increases anthropogenic heating of urban environments [47, 64]. The drivers of UHI present opportunities for a variety of mitigation strategies (Table 1) and mostly are passive design approaches. Santamouris [68] suggests the use of greenery and planted roofs, cool pavements, cool roofs and increase of urban albedo, are particularly effective urban mitigation techniques which could "*regulate the thermostat of cities*" (p 49). The architectural approach of the use of bioclimatic design strategies and passive and low-energy techniques can potentially accomplish about 80% reduction on building energy use consumption for heating, cooling and lighting [43]. Thus, leaving about 20% for designing efficient active mechanical and electrical systems and use of renewable energy sources both on- and off-site [3]. Many passive building design strategies can be adapted responding to location, size and purpose of buildings [54]. Good resources for the Australian context include: www.youhome.gov.au and www.builditbackgreen.org.

Addressing the adverse impacts of heatwaves on built environments means mitigating the effects by turning to factors that contribute to cooling demand. In Australia, building energy regulations primarily focuses on regulating the total maximum energy needed to heat and cool a building. Energy estimation in building rating and design relies on well-established thermal models which have been incorporated into the Nationwide House Energy Rating Scheme (NatHERS), www.nathers.gov.au, and integrated into the National Construction Code–Building Code of Australia (NCC-BCA) to establish minimum energy performance standards [80]. The primary role of the regulatory framework associated with the thermal model is to reduce

Table 1 Urban heat island mitigation strategies (adapted from Coutts et al. [17, pp. 42–43])

Approaches	How it works
Building design	Insulation and double glazing can reduce the need for heating and air conditioning which can improve energy efficiency
Increased albedo	Increasing the reflectivity of surfaces thereby limiting the heat transfer into buildings and heat storage
High thermal emittance surfaces	By covering roof materials with materials that reflect the near infrared can increase the albedo
Cool roofs	Use of materials having: high solar reflectance to reflect most of incident solar radiation during daytime, keeping the surface cooler and high thermal emittance allowing the materials to radiate away the heat stored in the structure, mainly during night-time
Green roofs	Reduce the heat transfer into the building while retaining water which encourages evapotranspiration. Green walls have similar effects
Outdoor landscaping	Vegetation planted in specific locations can block or limit the solar radiation reaching buildings and reduce the heat storage
Parkland and open space	Provides cooling for areas downwind, with vegetation encouraging evapotranspiration
Increase vegetation	Natural cooling system as it encourages evapotranspiration and energy is dissipated through latent heating instead of sensible heating
Street design	Widening streets as building heights increase allows a large sky view factor which helps with ventilation and cooling. Also, the orientation of the city affects the exposure to incoming solar radiation
Water-sensitive urban design	Increasing evaporation in the urban environment by retaining water
Energy efficiency	By using more efficient products, energy consumption can be reduced, minimising waste heat production
Mass transport	Increasing the use of public transport will reduce the number of vehicles that would contribute to CO_2 and anthropogenic heating

the annual energy needed to heat and cool the building [24]. Moore and Holdsworth, in Chapter "The Built Environment and Energy Efficiency in Australia: Current State of Play and Where to Next", provide a discussion on the regulatory frameworks governing residential energy efficiency in Australia. However, a multidisciplinary study completed in 2013, *A framework for the adaptation of Australian households to heatwaves* [65], found that focusing on peak cooling demand can have a sustained impact on peak electricity demand rather than the requirements of current building and air-conditioner regulations around energy usage. The proposed framework puts forward the case for regulatory changes to NatHERS energy rating tool in lieu of prescriptive measures to adapting house designs to climate change. With dwelling modifications, combined with enhanced regulation of air conditioners and household behaviour change strategies, Australian households can readily adapt to the impact of heatwaves and reduce the risk of heat-related deaths and household energy costs [65].

5 Why Should Building Professionals Be Concerned?

Beyond measures and strategies for the built environment, how is Australia placed in the commitments made in the Paris Agreement (2016) to reduce emissions by 2030 [22]? Critics and observers have been vocal that Australia's commitment to a 26–28% reduction of emissions by 2030 is one of the weakest emissions targets in the developed world [72].

Developments in attribution science [57] have allowed extreme event attribution to be more nuanced and by extension informed the way community is connecting climate change to the societal impacts of extreme weather. This shift in interest and understanding could have legal implications for policy and decision-makers with a duty of care to the community under a range of constitutional, common law or statutory rights [37]. Of great interest at present time is the discussion that due to the absence of enforceable commitments from governments, litigation may play an important role in reducing greenhouse gas emissions [48]. Certainly, government agencies are aware of the implications of human-induced extreme events, typically referred to as "*acts of God*", and are likely to inform claims and liability for damages. Interestingly, in 2011, the Australian Local Government Association (ALGA) commissioned a comprehensive review of the liability risks to local government that may arise because of climate change [4]. The report concludes that "*Councils must ensure they keep up to date with general climate change science and information related to mitigation and adaptation strategies... Councils will require localized information on impacts on which they can rely when making planning decisions and specialist advice on planning and engineering options for other aspects of adaptation*" [4, pp. 82–83].

In summary, the acknowledgement and understanding of the human impacts of climate change and the challenges these present to our cities, the built environment, and our way of life should be the impetus to exploring ongoing and new research and innovative developments that would pave directions to achieve a lower carbon society.

References

1. ABS (2012) 1301.0 Year Book Australia No. 92, 2012 Australian Bureau of Statistics (ABS), Canberra
2. AEMO (2017) System Event Report, New South Wales, 10 February 2017 (issued 22 February 2017)—Reviewable operating incident report for the National Electricity Market, Australian Energy Market Operator (AEMO)
3. Bainbridge D, Haggard K (2011) Passive solar architecture: heating, cooling, ventilation, daylighting and more using natural flows. Chelsea Green Publishing, White River Junction, Vermont
4. Baker & McKenzie (2011) Local council risk of liability in the face of climate change—resolving uncertainties a report for the Australian Local Government Association (22 July 2011), Australian Local Government Association (ALGA)
5. Bennett CM, Dear KBG, McMichael AJ (2014) Shifts in the seasonal distribution of deaths in Australia, 1968–2007. Int J Biometeorol 58(5):835–842
6. BoM (2013) Special Climate Statement 43—extreme heat in January 2013 Australian Government Bureau of Meteorology (BoM), viewed 28 Dec 2013. http://www.bom.gov.au/climate/current/statements/scs43e.pdf
7. BoM (2017) Special Climate Statement 61—exceptional heat in southeast Australia in early 2017 (issued 11 April 2017), Australian Government Bureau of Meteorology (BoM)
8. BoM (2017) Special Climate Statement 63—a prolonged warm spell in Tasmania and Victoria (issued 4 Dec 2017), Australian Government Bureau of Meteorology (BoM)
9. BoM (2018) Annual Climate Statement 2017 (10 January 2018), Australian Government Bureau of Meteorology (BoM), viewed 16 Jan 2018. http://www.bom.gov.au/climate/current/annual/aus/2017/
10. Cai W, Purich A, Cowan T, van Rensch P, Weller E (2014) Did climate change-induced rainfall trends contribute to the Australian millennium drought? J Clim 27(9):3145–3168
11. Climate Commission (2012) The critical decade: Victorian climate impacts and opportunities. Climate Commission Secretariat (Department of Climate Change and Energy Efficiency), Canberra, ACT
12. Climate Commission (2013) The Angry Summer, Climate Commission Secretariat (Department of Climate Change and Energy Efficiency), Canberra, ACT
13. Climate Commission (2013) Off the charts: extreme Australian summer heat, Climate Commission Secretariat (Department of Climate Change and Energy Efficiency), Canberra, ACT
14. Coates L, Haynes K, O'Brien J, McAneney J, de Oliveira FD (2014) Exploring 167 years of vulnerability: an examination of extreme heat events in Australia 1844–2010. Environ Sci Policy 42:33–44
15. Coutts A, Beringer J, Jimi S, Tapper N (2009) The urban heat island in Melbourne: drivers, spatial and temporal variability, and the vital role of stormwater
16. Coutts A, Beringer J, Tapper N (2007) Impact of increasing urban density on local climate: spatial and temporal variations in the surface energy balance in Melbourne, Australia. J Appl Meteorol Climatol 46(4):477–493
17. Coutts A, Beringer J, Tapper N (2010) Changing urban climate and CO_2 emissions: implications for the development of policies for sustainable cities. Urban Policy Res 28(1):27–47
18. CSIRO & BoM (2015) Climate change in Australia information for Australia's natural resource management regions: Technical Report, Commonwealth Scientific and Industrial Research Organisation (CSIRO) and the Australian Bureau of Meteorology (BoM)
19. CSIRO & BoM (2016) State of the climate 2016. Commonwealth Scientific and Industrial Research Organisation (CSIRO) and the Australian Bureau of Meteorology (BoM)
20. DCC (2009) Climate change risks to Australia's Coast—A First Pass National Assessment, Australian Government Department of Climate Change (DCC), Canberra, ACT
21. Deacon B, Doyle K (2017) Climate change increasing the intensity and impact of extreme weather, BOM tells emergency managers (04 September 2017). ABC News, viewed

6 Feb 2018. http://www.abc.net.au/news/2017-09-04/cyclone-and-extreme-weather-events-intensifying-bom-says/8869042
22. DEE (2016) Australia's 2030 emission reduction target, Australian Government Department of the Environment and Energy (DEE). http://www.environment.gov.au/climate-change/government/australias-emissions-reduction-target
23. DEE (2017) Australia's emissions projections 2017 (December 2017), Australian Government Department of the Environment and Energy, Canberra, ACT
24. Delsante A (2005) Is the new generation of building energy rating software up to the task?—A review of AccuRate', paper presented to ABCB Conference 'Building Australia's Future 2005, Surfer's Paradise
25. DIRD (2015) State of Australian Cities 2014-4015, Australian Government Department of Infrastructure and Regional Development (DIRD), Canberra, ACT
26. DIT (2011) State of Australian Cities 2011, Australian Government Department of Infrastructure and Transport (DIT), Major Cities Unit, Canberra, ACT
27. DIT (2013) State of Australian Cities 2013, Australian Government Department of Infrastructure and Transport (DIT), Canberra, ACT
28. Elrick-Barr CE, Smith TF, Preston BL, Thomsen DC, Baum S (2016) How are coastal households responding to climate change? Environ Sci Policy 63:177–186
29. Emmerson K, Hibberd M, Keywood M (2017) Australia State of the Environment 2016: Atmosphere, independent report to the Australian Government Minister for the Environment and Energy, Australian Government Department of the Environment and Energy, Canberra. Australian Government Department of Environment and Energy, Canberra, ACT
30. Erell E, Pearlmutter D, Williamson T (2011) Urban microclimate: designing the spaces between buildings. Taylor & Francis, New York
31. Freedman A (2013) Epic Heat, Wildfires Are Scorching Australian Landscape, viewed 15 Jan 2018. http://www.climatecentral.org/news/epic-heat-and-wildfires-torch-australia-15444
32. Goering L (2017) Rising heat: can we cool the risks of an invisible disaster? Thomson Reuters Foundation, 26 Sept 2017. http://news.trust.org/shorthand/risingheat/
33. Hannam P (2013) Temperatures off the charts as Australia turns deep purple. The Sydney Morning Herald, 8 Jan 2013
34. Hughes L, McMichael AJ (2011) The critical decade: climate change and health, Climate Commission Secretariat (Department of Climate Change and Energy Efficiency), Canberra, ACT
35. IPCC (2012) Managing the risks of extreme events and disasters to advance climate change adaptation. A Special Report of Working Groups I and II of the Intergovernmental Panel on Climate Change. Cambridge University Press, Cambridge
36. Jegasothy E, McGuire R, Nairn J, Fawcett R, Scalley B (2017) Extreme climatic conditions and health service utilisation across rural and metropolitan New South Wales. Int J Biometeorol 61(8):1359–1370
37. Jezard A (2017) This map shows extreme weather events in places influenced by human-made climate change (13 Dec 2017), World Economic Forum, viewed 6 Feb 2018. https://www.weforum.org/agenda/2017/12/this-map-shows-where-human-made-climate-change-influenced-extreme-weather-events/
38. Karoly D, Black MT, Grose MR, King A (2016) The roles of climate change and El Nino in the record low rainfall in October 2015 in Tasmania, Australia. Bull Am Meteor Soc 97(2):S217–S130
39. Kendal D, Baumann J (2016) The city of Melbourne's future urban forest—identifying vulnerability to future temperatures. The University of Melbourne and City of Melbourne, Melbourne, VIC
40. Kerslake DM (1972) The stress of hot environments, Monographs of the Pysiological Society No 29. Cambridge University Press, London
41. King A, Karoly D, Van Oldenborgh GJ, Hale M, Perkins-Kirkpatrick S (2017) Climate change's signature was writ large on Australia's crazy summer of 2017 (02 Mar 2017). The Conversation, viewed 6 Feb 2018, https://theconversation.com/climate-changes-signature-was-writ-large-on-australias-crazy-summer-of-2017-73854

42. Kjellstrom T (2000) Climate change, heat exposure and labour productivity. Epidemiology 11(4):S144
43. Lechner N (2014) Chapter 1—Heating, cooling and lighting as form-givers in architecture. In: Heating, cooling, lighting: sustainable design methods for architects, 4th edn, Wiley, Hoboken, pp 1–20
44. Lewis S, Perkins-Kirkpatrick S (2014) Human hands are all over Australia's hottest ever year (30 Sept 2014). The Conversation, viewed 26 Dec 2017. https://theconversation.com/human-hands-are-all-over-australias-hottest-ever-year-32267
45. Loughnan M, Tapper N, Thu Phan T, Lynch K, McInnes J (2013) A spatial vulnerability analysis of urban populations during extreme heat events in Australian capital cities, National Climate Change Adaptation Research Facility (NCCARF) and Monash University, Gold Coast
46. Lundgren-Kownacki K, Hornyanszky ED, Chu TA, Olsson JA, Becker P (2017) Challenges of using air conditioning in an increasingly hot climate. Int J Biometeorol
47. Lundgren K, Kjellstrom T (2013) Sustainability challenges from climate change and air conditioning use in urban areas. Sustainability 5(7):3116–3128
48. Marjanac S, Patton L, Thornton J (2017) Acts of God, human influence and litigation. Nat Geosci 10:616–619
49. McEvoy D, Ahmed I, Mullett J (2012) The impact of the 2009 heat wave on Melbourne's critical infrastructure. Local Environ 17(8):783–796
50. McInnes KL, Macadam I, O'Grady J (2009) The effect of climate change on extreme sea levels along Victoria's Coast—A Report the Victorian Department of Sustainability and Environment, CSIRO National Research Flagships Climate Adaptation
51. McKenzie A (2017) From 'angry summer' to 'weird winter': 2017 was riddled with extreme weather (28 Dec 2017). The Guardian, viewed 6 Feb 2018. https://www.theguardian.com/commentisfree/2017/dec/28/from-angry-summer-to-weird-winter-2017-was-riddled-with-extreme-weather
52. Oke TR (1982) The energetic basis of the urban heat island. Quart J Roy Meteorol Soc 108(455):1–24
53. Oke TR (1987) Boundary layer climates. 2nd edn, Routledge, London
54. Osmond P, Sharifi E (2017) Guide to Urban Cooling Strategies. Low Carbon Living CRC, Sydney, NSW
55. Parsons KC (2014) Human thermal environments: the effects of hot, moderate and cold environments on human health, comfort and performance, 3rd edn. CRC Press Taylor & Francis Ltd., London
56. Pearlman J (2018) Sydney swelters in heatwave, water usage at highest level in 15 years (8 Jan 2018). The Straits Times, viewed 11 Jan 2018. http://www.straitstimes.com/asia/australianz/weekend-heatwave-makes-sydney-the-hottest-city-on-earth
57. Pidcock R, Pearce R (2017) Mapped: How climate change affects extreme weather around the world, Carbon Brief Clear on Carbon, viewed 6 Feb 2018. https://www.carbonbrief.org/mapped-how-climate-change-affects-extreme-weather-around-the-world
58. PwC (2011) Protecting human health and safety during severe and extreme heat events: A national framework (A Report prepared for the Commonwealth Government), PriceWaterHouseCoopers Australia (PwC) and Australian Government Department of Climate Change and Energy Efficiency
59. Queensland University of Technology (2010) Impacts and adaptation response of infrastructure and communities to heatwaves: the southern Australian experience of 2009—A Report for the National Climate Change Adaptation Research Facility (NCCARF), National Climate Change Adaptation Research Facility (NCCARF), Melbourne, VIC
60. Rajagopalan P, Hien WN, Wai David CK (2008) Microclimatic modeling of the urban thermal environment of Singapore to mitigate urban heat island. Sol Energy 82(8):727–745
61. Robertshaw D (1981) Man in extreme environments, problems of the newborn and elderly. In: Cena KM, Clark JA (eds) Studies in environmental science: bioengineering, thermal physiology and comfort, vol 10. Elsevier Scientific Publishing Company, Amsterdam, pp 169–179

62. Rosier S, Dean S, Stuart S, Carey-Smith T, Black MT, Massey N (2015) Extreme rainfall in early July 2014 in Northland, New Zealand—Was there an anthropogenic influence? In: Herring SC, Hoerling MP, Peterson TC, Stott PA (eds) Explaining Extreme Events of 2014: From A Climate Perspective (Special Supplement to the Bulletin of the American Meteorological Society), vol 96. American Meteorological Society, Boulder, Colorado, pp S136–S140
63. Rowlinson S, YunyanJia A, Li B, ChuanjingJu C (2014) Management of climatic heat stress risk in construction: a review of practices, methodologies, and future research. Accid Anal Prev 66:187–198
64. Salamanca F, Georgescu M, Mahalov A, Moustaoui M, Wang M (2014) Anthropogenic heating of the urban environment due to air conditioning. J Geophys Res: Atmos 119(10):5949–5964
65. Saman W, Boland J, Pullen S, de Dear R, Soebarto V, Miller W, Pocock B, Belusko M, Bruno F, Whaley D, Pockett J, Bennetts H, Ridley B, Palmer J, Zuo J, Ma T, Chileshe N, Skinner N, Chapman J, Vujinovic N, Walsh M, Candido C, Deuble M (2013) A framework for adaptation of Australian households to heat waves—A Report for the National Climate Change Adaptation Research Facility (NCCARF), National Climate Change Adaptation Research Facility (NCCARF), Melbourne, VIC
66. Santamouris M (2014) On the energy impact of urban heat island and global warming on buildings. Energy Build 82:100–113
67. Santamouris M (2015) Analyzing the heat island magnitude and characteristics in one hundred Asian and Australian cities and regions. Sci Total Environ 512–513:582–598
68. Santamouris M (2015) Regulating the damaged thermostat of the cities—status, impacts and mitigation challenges. Energy Build 91:43–56
69. Santamouris M (2016) Cooling the buildings—past, present and future. Energy Build 128:617–638
70. Santamouris M, Haddad S, Fiorito F, Osmond P, Ding L, Prasad D, Zhai X, Wang R (2017) Urban heat island and overheating characteristics in Sydney, Australia. An analysis of multiyear measurements. Sustainability 9(5):712
71. Singh D, Horton DE, Tsiang M, Haugen M, Ashfaq M, Mei R, Rastogi D, Johnson NC, Charland A, Rajaratnam B, Diffenbaug NS (2014) Severe precipitation in Northern India in June 2013: Causes, Historical Context, and Changes in Probability. In: Herring SC, Hoerling MP, Peterson TC, Stott PA (eds) Explaining extreme events of 2013: from a climate perspective (Special Supplement to the Bulletin of the American Meteorological Society), vol 95. American Meteorological Society, Boulder, Colorado, pp S58–S66
72. Steffen W, Alexander D, Rice M (2017) Critical decade 2017: accelerating climate action. Climate Council of Australia Sydney, NSW
73. Steffen W, Hughes L, Perkins S (2014) Heatwaves: hotter, longer, more often. p 24, viewed 26 Dec 2017. http://www.climatecouncil.org.au/heatwaves-report
74. Steffen W, Rice M, Alexander D (2018) 2017: another record-breaking year for heat an extreme weather. Climate Council of Australia Sydney, NSW
75. Steffen W, Stock A, Alexander D, Rice M (2017) Angry summer 2016/17: Climate Change Super-charging Extreme Weather Climate Council of Australia Sydney, NSW
76. Stone B, Hess JJ, Frumkin H (2010) Urban form and extreme heat events: are sprawling cities more vulnerable to climate change than compact cities? Environ Health Perspect 118(10)
77. UK Met Office (2018) 2017: warmest year on record without El Niño, 18 Jan 2018 edn, UK Met Office, Exeter, United Kingdom, 18 Jan 2018. https://www.metoffice.gov.uk/news/releases/2018/2017-temperature-announcement
78. Wang SY, Hipps L, Gillies RR, Yoon J-H (2014) Probable causes of the abnormal ridge accompanying the 2013–2014 California drought: ENSO precursor and anthropogenic warming footprint. Geophys Res Lett 41(9):3220–3226
79. Watts N, Adger WN, Agnolucci P, Blackstock J, Byass P, Cai W, Chaytor S, Colbourn T, Collins M, Cooper A, Cox PM, Depledge J, Drummond P, Ekins P, Galaz V, Grace D, Graham H, Grubb M, Haines A, Hamilton I, Hunter A, Jiang X, Li M, Kelman I, Liang L, Lott M, Lowe R, Luo Y, Mace G, Maslin M, Nilsson M, Oreszczyn T, Pye S, Quinn T, Svensdotter M, Venevsky S, Warner K, Xu B, Yang J, Yin Y, Yu C, Zhang Q, Gong P, Montgomery H,

Costello A (2015) Health and climate change: policy responses to protect public health. Lancet 386(10006):1861–1914
80. Williamson T, Orkina N, Bennetts H (2009) A comparison of accredited second generation NatHERS software tools. Paper presented to 43rd Annual Conference of the Architectural Science Association, ANZAScA, 2009, University of Tasmania
81. WMO (2018) WMO confirms 2017 among the three warmest years on record, 18 Jan 2018 edn, World Meteorological Organization (WMO), Geneva, Switzerland, 18 Jan 2018. https://public.wmo.int/en/media/press-release/wmo-confirms-2017-among-three-warmest-years-record
82. Xiang J, Bi P, Pisaniello D, Hansen A (2014) Health impacts of workplace heat exposure: an epidemiological review. Ind Health 52(2):91–101

Dr. Mary Myla Andamon, B.Sc. Arch., Ph.D. is a lecturer and a building scientist with expertise in indoor and outdoor thermal comfort studies, urban microclimate and building environmental monitoring and performance assessments.

Mr. Andrew Carre, B.E. (Hons), B.Ec., M.E. is a lecturer and an energy engineer with expertise in life cycle assessment. He has considerable experience in research and client management.

Thermal Environments in the Construction Industry: A Critical Review of Heat Stress Assessment and Control Strategies

Ruwini Edirisinghe and Mary Myla Andamon

Abstract In the light of climate change predictions, the increasing number of hot days will cause a significant impact on public health, mortality rates, energy demand and economy of Australia. Heat is also an occupation hazard, which is a growing concern in many industries. Heat stress hazards can be categorized as clinical, human performance diminishing and accident causing. The risk can be exaggerated in certain industries, including the construction industry, due to specific environmental conditions, work characteristics and occupational settings. This chapter discusses the main problems and risks associated with heat stress, with a particular emphasis on the construction industry. Various heat stress indices and advances in the assessment of heat stress in recent years are discussed. Finally, this chapter discusses the strategies and controls that can be implemented to mitigate the impact of heat stress in the construction industry. Various acclimatization protocols, hydration, self-pacing and exposure time limits or temperature risk control regimes are discussed by analysing standards, guidelines and policies and practices. This chapter contributes to resolving a timely and strategic occupational hazard through a holistic view of the thermal environment in construction industry settings.

1 Introduction

Australia is the world's driest inhabited continent, and there is an extensive body of climatological evidence that the landscape and environments of the country differ significantly from those of other continents [32]. As discussed in Chapter "Urban Climates in the Transformation of Australian Cities", extreme heat events are projected to occur more frequently across the country and with increased severity, including in

R. Edirisinghe (✉) · M. M. Andamon
Sustainable Building Innovation Laboratory, School of Property Construction and Project Management, RMIT University, Melbourne, VIC, Australia
e-mail: ruwini.edrsinghe@rmit.edu.au

P. Rajagopalan et al. (eds.), *Energy Performance in the Australian Built Environment*, Green Energy and Technology, https://doi.org/10.1007/978-981-10-7880-4_3

all capital city locations [52] in Australia. This alerts us to the potential for exaggerated climatic heat stress in the future with potentially significant impact on public health, mortality rates, energy demand and economy [78, 82].

1.1 Heat Stress

Rowlinson et al. [77: p. 188] define heat stress as the 'heat load imposed on the human body, including environmental heat, metabolic heat and the thermal effect of clothing'. A dynamic heat balance between the body and its environment is required by the laws of thermodynamics; that is, heat transferred into the body and generated within the body must be balanced by heat outputs from the body in a dynamic manner [71 #7: p. 9, 72: p. 33, #125]. When the total of the heat transferred to the body and the heat generated within the body is greater than the heat output, body temperature rises.

The human thermal environment was originally fundamentally defined as consisting of four basic environmental factors (air temperature, humidity, radiant heat and wind speed) and two personal parameters of clothing and the amount of metabolic heat generated by physical activity [34] as discussed below. However, the literature suggests that other environmental factors such as barometric pressure are also influential [7, 13]. To present the environmental heat stress more accurately, it is important to go down to a microlevel. Urban microclimate (atmospheric) as well as clothing microclimate are crucial in this regard. Local microclimates can vary greatly based on such factors as topography, elevation, moisture, wind, soil and vegetation. Climatic heat stress ultimately occurs in the microclimate between the body and the clothing covering it, and thus the properties of the clothing are significant [34]. The insulation, permeability and water vapour resistance of the clothing [30] worn by the person significantly affect the heat exchange between the body and the environment. Metabolic heat is generated by activity. Posture, too, affects the heat exchange between the body and the environment. Hence, an understanding of the relevant activities and postures is crucial in analysing metabolic heat production. In addition to the fundamental factors needed to represent the minimum requirements for a conceptualization of heat stress conditions, a number of personal and external factors can influence the ability of an individual to dissipate excess heat [90].

1.2 Heat Stress Hazard

From a medical perspective, heat-induced illness covers a spectrum of disorders [53], ranging from minor to catastrophic based on the duration, severity and consequences of the risk. The spectrum of disorders covered by 'heat stress' includes physical fatigue, mental fatigue, heat-related chronic conditions and heat illnesses (where some are symptoms of heat stroke [10, 49]) in which the risk of heat stroke is catastrophic on the basis of its severe consequences.

Physical fatigue elevated by exposure to heat causes significant physical performance decrements [10]. Miller and Bates [57] flag the fact that physiological heat stress increases worker vulnerability to heat-related illness and decreases physical and cognitive performance and physical alertness [71]. Heat stress not only threatens survival, but also harms morale [13] and leads to deterioration of work efficiency and productivity [33] among workers.

Workplace accidents are more common in hot environments and are often associated with heat stress and dehydration [48]. Fatigue, which includes the deterioration of both physical and cognitive ability, is identified as one of the most important causes of construction accidents [20, 38]. In this regard, Edwards and Bowen [31] argue that heat stress presents a legal risk to organizations when accidents subsequently occur.

1.3 Construction Environments and Conditions

Conditions in construction sites vary with the geographic locations, types of construction and the stages of a project life cycle, particularly with the construction phases and stages of procurement of the projects. Depending on the site characteristics and construction work, the environmental parameters will amplify the effects of the ambient thermal conditions. In outdoor works such as civil works, concrete pouring and roofing [87], workers are more vulnerable to radiant heat from site characteristics, handling of construction materials and external building surfaces. For building works that offer shaded areas or enclosed conditions, environments will mostly be characterized as with high humidity with poor ventilation. The requirement of the use of construction personal protection equipment (PPE) also contributes to and exacerbates the construction workers' heat strain [58, 89].

While studies have addressed heat stress problem in general and among other vulnerable population, heat stress in the construction industry is emerging. The exposure and susceptibility of construction workers to heat stress have always been a challenge to the construction industry [88]. Various strategies and policies within the building industry sector have been implemented [14, 15, 16, 40, 64]. However, despite the existence of these guidelines and regulatory requirements, increased heat stress-related morbidity and mortality are widely reported [21, 50, 60, 73, 74, 77, 83, 84].

In this vacuum, Sect. 2 of this chapter discusses the heat stress risk in the construction industry. Risk assessment mechanisms used in the industry are also analysed. Section 4 presents the control regimes adopted by the industry. Section 5 presents the discussion, where gaps and recommendations are highlighted, and Sect. 6 concludes the chapter.

2 Heat Stress Risk in the Construction Industry

The construction industry has complex processes, and managing work health and safety in construction remains a wicked problem [29]. Globally, the construction industry records high accident rates and thus has been a priority industry for occupational health and safety improvements for decades [28, 42]. The construction industry is found to be more susceptible to heat stress than other industries due to its occupations' settings [84, 90]. Construction workers are vulnerable to heat stress factors (as set out above). These include: (i) direct exposure to climatic conditions (tropical/hot and humid or extreme environmental conditions, high radian heat loads or direct sun light); (ii) confined work environments and work environments near radiant heat sources; (iii) heat stress exacerbated by heavy industrial clothing and personal protective equipment (PPE); (iv) physically demanding work at a high metabolic rate; and (v) types of construction site (e.g. roads, buildings), project life cycles (indoor vs. outdoor) and construction activities which have a significant impact on climatic heat stress. The inherently dangerous construction industry is also inherently vulnerable to heat stress hazards.

3 Assessing Heat Stress

Physiological strain in the heat can be expressed in terms of the magnitude of core body temperature elevations, the volume of sweat lost (and the subsequent degree of dehydration if fluids are not fully replaced) [47] or loss of body mass through sweating [46], as well as, to a lesser extent, elevation in mean skin temperature [47]. Physical fatigue resulting from heat strain can be indicated by heart rate, oxygen uptake, blood pressure, respiration rate and/or perceived fatigue [10].

In the measurement and 'prediction' of the human response to thermal environments, thermal indices have proven useful in describing and assessing human thermal environments. Many studies have explored heat risk identification mechanisms [33] in general including the comfort literature reviews on indices [37, 39, 79, 86] Generally, these are categorized in three types [70]: those derived from heat balance equations and mathematical models that describe the behaviour of the human body in thermal environments (rational indices); those which are derived from experiments (empirical indices) and those based on measurements taken on simple instruments that respond to thermal environmental factors which also affect people (direct indices). An extensive discussion of these categories can be found in the ASHRAE Handbook Fundamentals [6]—application of heat stress indices and assessment methods in the construction industry.

The aim of this section is to review those mechanisms and indices that have been tried and tested in the construction industry or in equivalent industrial configurations. The industry sector is demarcated for the purposes of this section according to the definition of the Australian Construction and Mining Equipment Industry Group [14].

Some of these indices used in construction only look at one environmental parameter to assess heat stress [14], while others use more [67] (ISO 7243 1989 43]). Some indices assess or predict the heat stress, taking physiological [62] and/or perceptual [23] personal differences into account, including metabolic rates [12, 13]. Predictive models have been introduced that are based on multiple regression analysis of specific indices in conjunction with other factors. These indices and their application in the construction industry are discussed below. The indices are summarized in Table 1.

3.1 Single-Parameter Index

Some heat stress management protocols in the construction industry use a single environmental parameter, usually the ambient temperature, as the indicator of heat stress. For example, the Enterprise Bargaining Agreement (EBA) of the Australian Construction, Forestry, Mining and Energy Union (CFMEU) states that 35 °C is the limit for work [14]. The use of a single environmental condition as a heat stress indicator has been widely criticized in the literature [9, 35, 61]. A single environmental condition is an unreliable indicator upon which to base a decision to terminate work or reduce shift length. Bates [9] highlights the negative impact of such strategies on productivity and protection of workers' health.

3.2 Heat Index (NOAA)

The heat index system was developed by the US National Oceanographic and Atmospheric Administration (NOAA) [67]. It combines air temperature and relative humidity into a single value to indicate apparent hotness. Because this index factors in the evaporation of sweat from the skin through humidity, it is more reliable than using a single parameter. The US heat stress guidelines propose risk levels based on the 'heat index' and suggest protective measures based on several thresholds [68]. The heat stress guidelines also serve as the regulatory guidelines for the construction industry. However, the heat index does not take the effect of solar radiation into account when calculating the hotness of an environment and thus does not represent the environmental heat stress with sufficient accuracy.

Chan et al. [20] trialed heat index in the construction industry in Hong Kong with a sample of rebar workers. The model reported that many factors would affect human physiological response to heat stress triggering that the heat index as a function of temperature and relative humidity is only a rough indicator.

Table 1 Heat stress assessment methods used in the construction industry

Index	Parameters	Positives	Limitations
CFMEU single-parameter index	Ambient temperature	Simple, easy to use	Inaccurate, negative impact on productivity or health
NOAA heat index	Air temperature, relative humidity	More reliable than single parameter	Not represent heat stress with a sufficient accuracy
Natural WBGT index	Dry-bulb temperature, wet-bulb temperature, black globe temperature	Standardized, widely used, better representation of heat stress	Insensitive to wind speed, overestimation, weak correlation with strain parameters, effects of protective behaviours(self-pacing), age and BMI not covered, metabolic rate and clothing effects are ignored
PSI	Heat strain parameters	Accounts for protective behaviour, individual differences (age, BMI)	Difficult to measure on site, practical issues, calibration needed to individual factors for better accuracy
Perceptual heat stress	Human perception	High correlation with strain indices, generalization to climatic and working conditions is under-researched	Subjective measurements, not reliable scales of temperature
PHS model index	Temperature, humidity, globe temperature, air velocity, metabolic rate, clothing effect, body size, posture and wind direction	Best representation of related factors, model validated with a large sample	Not reliable with thick clothing, validated reliability in occupation setting is unavailable, complexities of use and taking measurements, negative impact on productivity
TWL	Dry-bulb, wet-bulb and globe temperatures, wind speed and atmospheric pressure	Works well with air movement	Assumption is clothing factor, only for self-paced workers
Regression model	VO2, minute ventilation (MV), respiratory exchange ratio (RER), metabolic equivalent (MET), energy expenditure (EE) heart rate, perceived exertion (RPE), WBGT	A model with a wide range of parameters	Practical issues in collecting parameters, small sample, not validated, regression models are less accurate, generalizability is an issue

3.3 *Wet-Bulb Globe Temperature (WBGT) Index*

The wet-bulb globe temperature (WBGT) index was originally developed based on the weighting of three environmental parameters: dry-bulb temperature, wet-bulb temperature and black globe temperature [85]. A shielded dry-bulb thermometer, a natural wet-bulb thermometer and a globe thermometer (a black globe heated by solar radiation) are used to capture these environmental parameters. The WBGT index was later standardized (ISO 7243 1989 43]) and is now widely used as a heat stress index throughout the world, with the guidelines and permissible threshold limits [2] in many industries, including construction and underground mining operations [9], being based on this measure.

Even though natural WBGT is a better index than the NOAA 'heat index' as a way of representing environmental hotness, its shortcomings have been widely recognized in the literature. Among the critiques are that: (i) it is insensitive to wind speed [13] and underestimates the effect of wind speed [58]; (ii) the WBGT index has overestimated the heat stress faced by subjects exposed to heat in many developing countries, such as China, India, Thailand and Dubai [10, 41]; (iii) there is weak correlation between WBGT and physiological strain parameters [23, 24, 56], and thus the index is unable to indicate the physiological responses in the body for a true representation of heat stress; (iv) it is unable to represent the effects of self-pacing, age and BMI [24]; and (v) it is unable to measure the effect of the other non-environmental heat stress factors, namely metabolic rate and the clothing effect [71].

As widely recognized by the literature, the environmental indices on their own have limitations in reliably assessing/predicting heat stress due to the individual differences and complexity of the variables associated with the differences.

3.4 *Physiological Strain Index (PSI)*

The physiological strain index considers the individual heat strain parameters in contrast to the environmental parameters discussed above. PSI is based on heart rate and core temperature measurements and is represented on a scale from zero to 10 [62]. PSI can take into account parameters such as age and BMI and can also account for protective behaviour (e.g. self-pacing) [58].

3.5 *Perceptual Heat Stress*

The thermal sensation of hot or cold is psychological phenomena, and although there are physiological mechanisms in the body which respond to temperature, thermal sensation depends upon such things as previous experience, individual differences and rates of change of temperature. Humans are therefore not good temperature-measuring instruments and cannot provide reliable scales of temperature [71].

However, Dehghan et al. [23] argue that the body response to heat stress is known as the strain by which physiological and psychological parameters are measuring. They further argue that the physiological strain index together with observational–perceptual method shows a higher correlation with PSI than WBGT [8, 23]. Supportively, Chan and Yang [19] recently validated perceptual strain index (PeSI) developed by Tikuisis et al. [80] in the construction industry and found that PeSI is sensitive to the variants of WBGT and RHR and changes in the same general manner as PSI. Generalizability of the PeSI in various climatic and working environments is yet to be reported.

3.6 Predictable Heat Strain Model Index (ISO 7933:2004)

The predictable heat strain model [55] was developed and validated as a collaboration between eight major European laboratories, to ground the originally defined 'required sweat rate' model in ISO 7933:1989 [44] in a practical manner. For example, ISO 7933 is able to predict the sweat rate for constant climatic and working conditions. The PHI model was standardized as ISO 7933:2004 [45].

The PHS model takes air temperature, humidity, globe temperature, air velocity, metabolic rate, clothing effect, body size, posture and wind direction as inputs and provides a detailed analysis of the working conditions with predicted and required parameters such as the sweat rate, evaporative heat flow, the skin wettedness and the rectal temperatures. Despite the fact that which temperature more accurately describes the body thermal state is still an open question, in occupational terms rectal temperature is often assumed as being representative of the thermal state of an individual and is used in ISO 7933:2004 [75]. ISO 7933:2004 [45] suggests maximum allowable exposure duration. Rowlinson and Jia [76] applied the PHS model in the Hong Kong construction industry. This study was replicated by Lundgren et al. [54] in their Indian sample. Rowlinson and Jia [76] estimated metabolic rates based on heart rates which might have overestimated the metabolic rate. Even though it is protective of workers, can introduce unintended productivity issues. In addition, the study used tympanic temperature at the start instead of rectal temperature which has affected the accuracy defined in the ISO 7993:2004. Moreover, Rowlinson and Jia [76] assumed typical summer clothing ensemble and the effect of safety helmet was ignored which has a significant influence on body thermal state. Wang et al. [81] proved the lack of reliability of PHS index when it is used with thick protective clothing. It is paramount to evaluate the accuracy of the model with the actual parameters of clothing assembly. Hence, validated evidence on reliability of this method in various local settings is yet to be reported.

ISO 7933:2004 is often criticized when introduced into a workplace where large number of workers are involved, for complexities of use [58] and potential interruptions to the working environment and activities [76] which can have a negative consequence on productivity rather than a positive one.

3.7 Thermal Work Limit (TWL)

Brake and Bates [13] propose a heat index combined with data on environmental, metabolic and clothing factors called the 'thermal work limit' for workers who are well educated about working in heat, have control over their work rate, are healthy and are well hydrated. TWL uses five environmental variables: dry-bulb, wet-bulb, and globe temperatures, wind speed and atmospheric pressure and accommodates for clothing factors. As the equations used to derive heat transfer rate through clothing are not valid for subjects in encapsulating protective clothing (EPC), TWL cannot be assumed to be valid where impermeable clothing is used. TWL is particularly suitable when there is significant cooling related to air movement.

TWL model was validated in the Australian mining industry [58] and subsequently included in heat stress management guidelines and standards in Australia [5, 11, 25] and Abu Dhabi [27].

Dehydration status of construction workers was assessed using urine specific gravity (USG) measurements to indicate the absolute hydration status of the body in Australia [57], UAE [10] and Iran [9, 61]. All the studies found that the USG could be used as an indicator of thermal heat stress. While Bates and Schneider [10] found that use of WBGT as a thermal index is inappropriate for the study sample of 22 participants studies over 3 days, however, TWL was found to be a valuable index. In contrast, Farshad et al. [35] concluded that both GBWT and TWL were good indicators of heat stress in Iran climate but TWL has merit due to its based-on-required-intervention classifications.

3.8 Multiple Regression Analysis-Based Heat Stress Models

Chan et al. [17] conducted a study in Hong Kong construction industry. Prior to the experiments, they collected demographic data including age, behavioural habits (smoking, drinking) and other personal information together with body weight, percentage of body fat (PBF), resting heart rate and blood pressure. During the experiments, physiological data such as VO2, minute ventilation (MV), respiratory exchange ratio (RER), metabolic equivalent (MET), energy expenditure (EE) heart rate were monitored every five second. Rating of perceived exertion (RPE) was also recorded every five minutes. Environmental data on ambient dry-bulb temperature, natural wet-bulb temperature, globe temperature and relative humidity were also collected to calculate WBGT.

A similar study was conducted with a sample of rebar workers by Chan et al. [18] using TWL. These heat stress models were derived to predict workers' physiological responses, different metrological factors, work-related factors and personal factors based on multiple regression analyses. As Rowlinson et al. [77] argue complexity of the factors affecting heat stress is beyond the predictive power of multiple regression models. Hence, generalizability of the model to varying trades with in construction industry and to climatic conditions is yet to be reported.

4 Heat Stress Control Regimes

The National Institute for Occupational Safety and Health [63] provides standards of working practices to address hot environments. The three categories of standard are based on the recommendations and control methods NIOSH published in 1986, and the categories are (i) engineering controls; (ii) administrative controls; and (iii) personal protective clothing and auxiliary body cooling. When applying engineering controls for heat stress is not practical or sufficient, administrative strategies can be implemented to control heat risk. A significant number of research studies on heat stress control in construction using such administrative controls have been reported in the literature.

The NIOSH standard [63] recommends education and training on heat stress, for both workers and management. Training programmes at both levels should include recognition of heat stress (i) signs and symptoms, (ii) causes, (iii) the impact of PPE, (iv) the effect of non-occupational factors (drugs, alcohol and obesity), (v) the importance of acclimatizing, (vi) procedures for responding to symptoms and (vii) the importance of hydration. In addition to self-awareness, the NIOSH standard also emphasizes the importance of enhancing heat tolerance. Regular medical programmes and health screening methods are also recommended to capture workers' histories of heat illness and to monitor heat tolerance. Other control measures are discussed below.

4.1 Acclimatization Protocols

Human populations are acclimatized to their local climates, in physiological, behavioural and cultural terms. Stimulation of human heat-adaptive mechanisms can increase the capacity to tolerate work in heat [13, 63]. Even though a simple and practical measure of acclimatization is not available, some robust protocols have been designed to increase the ability of workers to work in hot environments. Heat acclimatization can usually be induced in 7 to 14 days of exposure at a hot job [3, 22, 66].

4.2 Hydration

Miller and Bates [57] studied fluid balance by monitoring fluid intake and hydration levels through urine specific gravity of Australian mining workers. They argue that the creation of a culture of hydration awareness in a workforce is an important component of a heat stress risk management strategy for workers. Supportively, Montazer et al. [61] argue that heat stress management without considering the real hydration status of workers is inadequate. However, self-hydration without an active campaign of the kind recommended by Miller and Bates [57] was challenged by the findings of Montazer et al. [61], in which they reported that the USG level of workers increased during midday work because the workers were asked to drink a specific volume of water during their work.

4.3 Self-pacing

Self-paced workers are defined as those who can and do regulate their own work rate, are not subject to excessive peer or supervisor pressure or monetary incentives and are well educated about the issues of working in heat and the importance of self-pacing [13: p. 176]. Brake and Bates [13] argue that for heat stress risk management to be effective, self-pacing should be formally incorporated in a protocol mandating workers to self-pace and supported by supervisors and management. In addition, the study of Australian mining workers [12] found that self-pacing occurs among well-informed workers. Supportively, Bates and Schneider [10], with a sample of UAE construction workers, found that people can work, without adverse physiological effects, in hot conditions if they are provided with the appropriate fluids and are allowed to self-pace. A further study in the UAE found that uneducated workers also regulate their workload in thermally stressful conditions [59].

Combining all three control strategies, Miller et al. [59] argue that well-hydrated, acclimatized workers who are permitted to self-pace may safely continue working under fluctuating harsh environmental conditions.

4.4 Limiting Exposure Time or Temperature

Guidelines for work–rest schedules and practical intervention levels and protocols are discussed below.

4.5 Construction, Forestry, Mining and Energy Union (CFMEU)

The CFMEU hot weather policy [14] recommends stopping work and leaving the site when air temperature reaches 35 °C. The agreement also states that at temperatures below 35 °C, workers are to be relocated out of direct sunlight where the work environment creates a serious risk to their health and safety. These serious risks include: (i) radiant heat from particular surfaces like bondeck, roofing; (ii) sun glare; and (iii) the type of work being performed.

4.6 ACGIH Threshold Limit Values (TLV) and Action Limit (AL) for Thermal Stress

The ACGIH [3] suggests threshold limit values (TLVs) for thermal stress. The objective of the TLV system is to maintain core body temperature within +1oC of the normal value (37 °C). TLV for heat-acclimatized, hydrated, un-medicated, healthy workers and action limit (AL) for un-acclimatized workers are expressed as time-weighted average (TWA) exposure for an eight-hour workday and 40-h (five-day) workweek. The effective wet-bulb globe temperature (WBGT) is derived based on the measured WBGT (then environmental index), plus the clothing adjustment factor (where clothing adjustment factor cannot be added for multiple layers). Empirical data are used to estimate metabolic rate. The time-weighted average of the effective WBGT accounts for the metabolic rate based on the work–rest regimen.

4.7 NIOSH Recommended Alert Limits (RALs) and Recommended Exposure Limits (RELs)

NIOSH is the US federal agency responsible for promoting occupational safety and health and recommends limiting the level of health risk associated with the total heat load imposed on a worker in a hot environment [63]. Recommended alert limits (RALs) are for un-acclimatized workers, whereas recommended exposure limits (RELs) are for acclimatized healthy workers, where the workers should be able to tolerate the heat stress without incurring adverse effects. Estimates of both environmental and metabolic heat are expressed as one-hour time-weighted averages (TWAs), as described by the ACGIH [4]. However, these limits are applicable to workers wearing the conventional one-layer work clothing ensemble. RAL and REL estimations are based on empirical data [26, 51, 63].

4.8 TWL-Based Interventions

The thermal work limit [13] gives a limiting (or maximum) sustainable metabolic rate (LMR) that acclimatized individuals can maintain in a specific thermal environment, in the form of a safe deep body core temperature (<38.2 °C or 100.8 °F) and a sweat rate (<1.2 kg or 2.6 lb/h). TWL predicts the limiting work rates under given environmental conditions, and interventions are recommended accordingly, such as withdrawal (if TWL value <115 W/m^2), buffer (115–140 W/m^2), acclimatization (141–220 W/m^2) and unrestricted work (>220 W/m^2).

Montazer et al. [61] found a strong correlation between TWL and USG, and a significant difference between the control group and the group exposed to heat. The maximum TWL levels were observed in the middle of the work shift. Bates and Schneider [10] indicated that with interventions to encourage fluid intake, self-paced construction workers were able to work in extreme temperatures, often in excess of 45 °C, with no evidence of physiological strain as assessed from working heart rates and aural temperature readings.

4.9 PHS Model (ISO 7933 2004)

The PHS model [45] predicts the maximum allowable exposure duration and provides a sensitivity analysis for testing the impact of specific parameters, including environmental heat, clothing effect and metabolic heat. Rowlinson and Jia [76] found that environmental thresholds computed by the PHS model, based on their sample, are 2–3 °C WBGT higher than the equivalent TLVs and thus argue that these thresholds can be used for initial screening but not as action-triggering thresholds.

Rowlinson and Jia [76] developed localized threshold-based guidelines for practical implementation using a number of the guidelines discussed above. They used the PHS model to develop a tool to facilitate managerial decision-making on an optimized work–rest regimen for paced work. Further, they used a TWL model-based limiting metabolic rate (LMR) to develop a tool to enable workers' self-regulation during self-paced work. The recovery time following a period of paced work was calculated using TWA [2].

5 Discussion

A preventive measure called the heat alert programme by NIOSH recommends establishing a heat alert committee during hot seasons to declare heat alerts and execute appropriate actions accordingly. Note that maintaining an effective heat alert committee is quite a resource-intensive administrative process. In construction settings, stakeholder groups come together for a short period of time [36] to complete a job and disband upon project completion, often without forming long-term working relationships beyond the scope of the project [65]. Due to the dynamic nature of the industry and this project-based group cohesiveness, the practical aspects of ensuring regular medical screenings and setting up heat alert committees are challenging compared to industries with regular permanent workforces.

In addition to self-awareness, the NIOSH standard also emphasizes the importance of enhancing heat tolerance, which has been trialed in other industries such as defence [22, 66]. Acclimatization protocols are practical for the construction industry to adopt. We recommend embedding them in ongoing heat stress training or occupational health and safety programmes.

Intervention strategies, such as those developed by the American Conference of Governmental Industrial Hygienists [3], and the thermal work limit (TWL) [13], specify threshold values for acclimatized and non-acclimatized workers. According to the ACGIH [3], a worker is considered acclimatized when they have a recent history of heat stress exposure of at least two continuous hours over between 5 of the last 7 days and 10 of the last 14 days. Nevertheless, evidence as to whether workers exposed intermittently to various lengths and amounts of heat stress during their jobs develop heat acclimatization similar to that achieved by continuously exposed workers is yet to be reported.

Similarly to acclimatization, the electrolyte and water balance problems of intermittently heat-exposed workers in comparison with continuously heat-exposed workers are still unknown.

Regardless of the practical implications, the standards also recommend decreasing work time and increasing workforce size to reduce metabolic heat load, in addition to introducing mechanization. Productivity remains a subject of debate in the Australian construction industry, which is a serial productivity under-performer [1, 69]. Because of this, the implementation of strategies that consume more resources can be challenging. Nevertheless, in financial year 2013, an annual loss of $6 billion worth of labour productivity due to climatic heat stress was reported in Australia [91], which amounts to between 0.33 and 0.47% of GDP. This would seem to be sufficient to justify extra investment in heat stress mitigation.

6 Conclusions

This chapter has discussed the human thermal environment in a significant occupational setting. It presented an assessment of the heat stress hazards in construction, as well as risk assessment and risk control regimes. Research and practice gaps and recommendations were derived. It is of paramount importance that urban microclimates are measured in order to assess the degree of environmental heat stress in construction, and Chapter "Urban Climate in the Transformation of Australian Cities" provides more detail about these microclimates in the context of climate change. It is vital to consider the requirements of vulnerable populations, such as construction workers, given their specific occupational settings, which can exacerbate heat stress. The importance of addressing the requirements of other vulnerable population such as low-income population and school children is discussed in Chapters "Low-Energy Housing as a Means of Improved Social Housing: Benefits, Challenges and Opportunities" and "Indoor Environmental Quality of Preparatory to Year 12 (P-12) Educational Facilities in Australia: Challenges and Prospects", respectively.

References

1. ABS ABOS (2015) Estimates of industry multifactor productivity, 2014–15 [Online]. Available: http://www.abs.gov.au/ausstats/abs@.nsf/mf/5260.0.55.002 [Accessed]
2. ACGIH (2009) Heat stress and strain: TLV(R) physical agents, 7th edn. documentation. ACGIH Signature Publications, Cincinnati
3. ACGIH (2012) TLVs and BEIs: threshold limit values for chemical substances and physical agents and biological exposure indices. American Conference of Governmental Industrial Hygienists, Cincinnati
4. ACGIH (2014) TLVs and BEIs: threshold limit values for chemical substances and physical agents and biological exposure indices. American Conference of Governmental Industrial Hygienists, Cincinnati
5. AIOH AIOOH (2003) A guide to managing heat stress. Australian Institute of Occupational Hygienists
6. ASHRAE (2017a) ASHRAE Handbook—fundamentals. American Society of Heating, Refrigerating and Air-Conditioning Engineers, Inc. (ASHRAE), Atlanta
7. ASHRAE (2017b) Thermal comfort (Chapter 9). In: ASHRAE handbook—fundamentals, SI edn. American Society of Heating, Refrigerating and Air-Conditioning Engineers, Inc. (ASHRAE), Atlanta
8. Baker J, Grice J, Roby L, Matthews C (2000) Cardiorespiratory and thermoregulatory response of working in firefighter protective clothing in a temperate environment. Ergonomics 43:1350–1358
9. Bates G (2002) Empirical validation of a new heat stress index. J Occup Health Saf Aust N Z 18:145–154
10. Bates GP, Schneider J (2008) Hydration status and physiological workload of UAE construction workers: a prospective longitudinal observational study. J Occup Med Toxicol 3:21
11. Bell S (2012) Commissioner for mine safety and health: Queensland mines inspectorate annual performance report 2011–12. Queensland Mines Inspectorate, Department of Natural Resources and Mines, State of Queensland, Queensland, Australia
12. Brake DJ, Bates GP (2002a) Deep body core temperatures in industrial workers under thermal stress. J Occup Environ Med 44:125–135
13. Brake DJ, Bates GP (2002b) Limiting metabolic rate (thermal work limit) as an index of thermal stress. Appl Occup Environ Hyg 17:176–186
14. CFMEU (2016) 35 °C, that's enough—CFMEU hot weather policy [Online]. Available: https://vic.cfmeu.org.au/news/35%C2%B0-c-that%E2%80%99s-enough-cfmeu-hot-weather-policy [Accessed]
15. CFMEU Queensland and Northern Territory (2015) Heat stress policy. Construction, Forestry, Mining and Energy Union (Queensland and North Territory Branch)
16. CSAO (2000) Heat stress: guidelines for recognition, assessment, and control in construction. Construction Safety Association of Ontario, Etobicoke
17. Chan AP, Yam MC, Chung JW, Yi W (2012) Developing a heat stress model for construction workers. J Facil Manage 10:59–74
18. Chan AP, Yi W, Chan DW, Wong DP (2013) Using the thermal work limit as an environmental determinant of heat stress for construction workers. J Manage Eng 29:414–423

19. Chan AP, Yang Y (2016) Practical on-site measurement of heat strain with the use of a perceptual strain index. Int Arch Occup Environ Health 89(2):299–306
20. Chan APC, Wong FKW, Yam MCH, Chan DMW, Mok ECM, Shea GYK, Dingsdag DA (2011) A research framework for assessing the effects of heat stress on construction workers. In: 6th International Structural Engineering and Construction Conference, ISEC06—Modern Methods and Advances in Structural Engineering and Construction. Zurich
21. Chi S, Han S (2013) Analyses of systems theory for construction accident prevention with specific reference to OSHA accident reports. Int J Project Manage 31:1027–1041
22. DOD (2003) Technical bulletin: heat stress control and heat casualty management. TB MED 507/AFPAM 48-152 (I). DOD, Washington
23. Dehghan H, Habibi E, Khodarahmi B, Ha HY, Hasanzadeh A (2013) The relationship between observational–perceptual heat strain evaluation method and environmental/physiological indices in warm workplace. Pak J Med Sci 2013(29):358–362
24. Dehghan H, Mortazavi SB, Jafari MJ, Maracy MR (2012) Evaluation of wet bulb globe temperature index for estimation of heat strain in hot/humid conditions in the Persian Gulf. J Res Med Sci. Off J Isfahan Univ Med Sci 17:1108
25. Department of Natural Resources and Mines (2012) Risk management of heat exposure in mining, mines safety bulletin [Online]. Available: https://www.dnrm.qld.gov.au/mining/safety-and-health/alerts-bulletins-search1/alerts-bulletins/mines-safety/risk-mgt-heat-exposure-mining [Accessed]
26. Dukes-Dobos FN, Henschel A (1973) Development of permissible heat exposure limits for occupational work. ASHRAE, J Am Soc Heating Refrigerating Air-Conditioning Eng 57–62
27. EHS Center (2012) Technical guideline: safety in the heat (Abu Dhabi EHSMS Regulatory Framework). Abu Dhabi Environment. Health and Safety Center, UAE
28. Edirisinghe R, Lingard H (2016) Exploring the potential for the use of video to communicate safety information to construction workers: case studies of organizational use. Construction Manage Econ 34:366–376
29. Edirisinghe R, Stranieri A, Blismas N (2016) Information visualisation for the wicked problem of safe construction design. Archit Eng Des Manage 12:296–310
30. Edirisinghe R, Jadhav A (2017) Is the smart safety vest a brutal innovation? Evaluation of microclimate performance using a thermal manikin. In: Chan PW, Neilson CJ (eds) 33rd Annual ARCOM Conference, 4–6 September 2017, Cambridge, UK. Association of Researchers in Construction Management, pp 734–744
31. Edwards PJ, Bowen PA (1998) Risk and risk management in construction: a reviewand future directions for research. Eng Constr Archit Manage 5:339–349
32. Emmerson K, Hibberd M, Keywood M (2017) Australia State of the Environment 2016: Atmosphere, independent report to the Australian Government Minister for the Environment and Energy, Australian Government Department of the Environment and Energy, Canberra. Australian Government Department of Environment and Energy, Canberra
33. Epstein Y, Moran DS (2006) Thermal comfort and the heat stress indices. Ind Health 44:388–398
34. Fanger PO (1970) Thermal comfort. Analysis and applications in environmental engineering
35. Farshad A, Montazer S, Monazzam MR, Eyvazlou M, Mirkazemi R (2014) Heat stress level among construction workers. Iran J Public Health 43:492
36. Fischer M, Waugh L, Axworthy A (1998) IT support of single project, multi-project and industry-wide integration. Comput Ind 35:31–45
37. Gagge AP (1981) Rational temperature indices of thermal comfort. In: Cena KM, Clark JA (eds) Studies in environmental science: bioengineering, thermal physiology and comfort. Elsevier Scientific Publishing Company, Amsterdam
38. Garrett JW, Teizer J (2009) Human factors analysis classification system relating to human error awareness taxonomy in construction safety. J Constr Eng Manage 135:754–763
39. Goldman RF (1988) Standards for human exposure to heat. In: Mekjavic IB, Banister EW, Morrison JB (eds) Environmental ergonomics: sustaining human performance in harsh environments. Taylor & Francis, Ltd., London

40. HSE HASE (2017) Measuring heat stress [Online]. Available: http://www.hse.gov.uk/temperature/assets/docs/heat-stress-checklist.pdf [Accessed]
41. Holmer I (2010) Climate change and occupational heat stress: methods for assessment. Glob Health Action 2010:5719
42. Hoonakker P, Loushine T, Carayon P, Kallman J, Kapp A, Smith MJ (2005) The effect of safety initiatives on safety performance: a longitudinal study. Appl Ergon 36:461–469
43. ISO 7243 (1989) Hot environments—estimation of the heat stress on working man, based on the WBGT-Index (Wet Bulb Globe Temperature). International Standard Organisation, Geneva
44. ISO 7933 (1989) Hot environments—analytical determination and interpretation of heat stress using calculation of the required sweat rate. International Standard, Organisation, Geneva
45. ISO 7933 (2004) Ergonomics of the thermal environment analytical determination and interpretation of heat stress using calculation of the predicted heat strain. International Standard, Organisation, Geneva
46. ISO 9886 (2004) Evaluation of thermal strain by physiological measurements. International Organisation for Standardisation, Geneva
47. Jay O, Brotherhood JR (2016) Occupational heat stress in Australian workplaces. Temperature 3:394–411
48. Kenefick RW, Hazzard MP, Armstrong LE (2003) Minor heat illnesses. In: Exertional heat illnesses. Human Kinetics Publishers Inc., USA
49. Kilbourne EM (1997) Public health consequences of disasters. In: Noji E (ed). Oxford University Press, New York
50. Kovats RS, Hajat S (2008) Heat stress and public health: a critical review. Annu Rev Public Health 29:41–55
51. Lind AR (1963) A physiological criterion for setting thermal environmental limits for everyday work. J Appl Physiol 18:51–60
52. Loughnan ME, Tapper NJ, Phan T, Lynch K, Mcinnes JA (2013) A spatial vulnerability analysis of urban populations during extreme heat events in Australian capital cities. National Climate Change Adaptation Research Facility (NCCARF), Gold Coast
53. Lugo-Amador NM, Rothenhaus T, Moyer P (2004) Heat-related illness. Emerg Med Clin N Am 22(2):315–327
54. Lundgren K, Kuklane K, Venugopal V (2014) Occupational heat stress and associated productivity loss estimation using the PHS model (ISO 7933): a case study from workplaces in Chennai, India. Glob Health Action 7:25283
55. Malchaire J, Piette A, Kampmann B, Mehnert P, Gebhardt H, Havenith G, Griefahn B (2001) Development and validation of the predicted heat strain model. Ann Occup Hyg 45:123–135
56. Malchaire J, Gebhardt H, Piette A (1999) Strategy for evaluation and prevention of risk due to work in thermal environments. Ann Occup Hyg 43:367–376
57. Miller V, Bates G (2007) Hydration of outdoor workers in north-west Australia. J Occup Health Saf Aust N Z 23:79
58. Miller VS, Bates GP (2007) The thermal work limit is a simple reliable heat index for the protection of workers in thermally stressful environments. Ann Occup Hyg 51:553–561
59. Miller V, Bates G, Schneider JD, Thomsen J (2011) Self-pacing as a protective mechanism against the effects of heat stress. Ann Occup Hyg 55:548–555
60. Mining Australia (2013) Santos sub-contractor dies of suspected heatstroke [Online]. Available: https://www.australianmining.com.au/news/santos-sub-contractor-dies-of-suspected-heatstroke/ [Accessed]
61. Montazer S, Farshad AA, Monazzam MR, Eyvazlou M, Yaraghi AAS, Mirkazemi R (2013) Assessment of construction workers' hydration status using urine specific gravity. Int J Occup Med Environ Health 26:762
62. Moran DS, Shitzer A, Pandolf KB (1998) A physiological strain index to evaluate heat stress. Am J Physiol 275:R129–R134
63. NIOSH (2016) NIOSH criteria for a recommended standard: occupational exposure to heat and hot environments. In: Jacklitsch B, Williams WJ, Musolin K, Coca A, KIM J-H, Turner N (eds) DHHS (NIOSH) Publication 2016–106. Department of Health and Human Services,

Centers for Disease Control and Prevention, National Institute for Occupational Safety and Health, Cincinnati
64. NZ Occupational Safety & Health Service (1997) What you need to know about temperature in places of work. NZ Occupational Safety & Health Service and Department of Labour, Wellington
65. Navon R, Sacks R (2007) Assessing research issues in Automated Project Performance control (APPC), Autom Constr 16(2007): 474–484
66. Navy Environmental Health Center (2007) Prevention and treatment of heat and cold stress injuries. Technical Manual NEHC-TM-OEM 6260.6A. [Online]. Navy Environmental Health Center, Bureau of Medicine and Surgery, Portsmouth. Available: http://www.public.navy.mil/surfor/Documents/6260_6A_NMCPHC_TM.pdf [Accessed]
67. OSHA, OSAHA (2016a) About the heat index [Online]. Available: https://www.osha.gov/SLTC/heatillness/heat_index/about.html [Accessed]
68. OSHA, OSAHA (2016b) Protective measures to take at each risk level [Online]. Available: https://www.osha.gov/SLTC/heatillness/heat_index/protective_low.html [Accessed]
69. PWC (2013) Productivity Score Card
70. Parsons KC (2000) Environmental ergonomics: a review of principles, methods and models. Appl Ergon 31:581–594
71. Parsons K (1993) Human thermal environments: the effects of hot, moderate, and cold environments on human health, comfort, and performance. Taylor & Francis, London
72. Parsons KC (2014) Human thermal environments: the effects of hot, moderate and cold environments on human health, comfort and performance. CRC Press Taylor & Francis Ltd., London
73. Queensland WC (2016) A current affair melting man [Online]. Available: https://www.worksafe.qld.gov.au/forms-and-resources/films/melting-man [Accessed]
74. Rameezdeen R, Elmualim A (2017) The impact of heat waves on occurrence and severity of construction accidents. Int J Environ Res Public Health 14:70
75. Rodrigues JM, Oliveira AVM, Gaspar AR, Raimundo AM, Quintela DA (2016) Working conditions in the ceramic industry: assessment of the heat exposure with the Predicted Heat Strain (PHS) index. Occup Saf Hyg IV 249
76. Rowlinson S, Jia YA (2014) Application of the predicted heat strain model in development of localized, threshold-based heat stress management guidelines for the construction industry. Oxf J—Ann Occup Hyg 58:326–339
77. Rowlinson S, Yunyanjia A, Li B, Chuanjingju C (2014) Management of climatic heat stress risk in construction: a review of practices, methodologies, and future research. Accid Anal Prev 66:187–198
78. Steffen W, Stock A, Alexander D, Rice M (2017) Angry summer 2016/17: climate change super-charging extreme weather Sydney. Climate Council of Australia, NSW
79. Thornley DL (1969) Criteria for thermal comfort. In: Sherratt AFC (ed) Air conditioning system design for buildings. Elsevier Publishing Company Ltd., Amsterdam
80. Tikuisis P, McLellan TM, Selkirk G (2002) Perceptual versus physiological heat strain during exercise-heat stress. Med Sci Sports Exerc 34:1454–1461
81. Wang F, Kuklane K, Gao C, Holmér I (2010) Can the PHS model (ISO7933) predict reasonable thermophysiological responses while wearing protective clothing in hot environments? Physiol Meas 32(2): 239
82. Williamson TJ, Grant E, Hansen A, Pisaniello D, Andamon MM (2009) An investigation of potential health benefits from increasing energy efficiency stringency requirements—Building Code of Australia volumes one and two. Adelaide Research and Innovation Pty Ltd., The University of Adelaide, Adelaide
83. Xiang J, Bi P, Pisaniello D, Hansen A (2014) Health impacts of workplace heat exposure: an epidemiological review. Ind Health 52:91–101
84. Xiang J, Bi P, Pisaniello D, Hansen A, Sullivan T (2014) Association between high temperature and work-related injuries in Adelaide, South Australia, 2001–2010. Occup Environ Med 71:246–252

85. Yaglou CP, Minaed D (1957) Control of heat casualties at military training centers. Arch Ind Health 16:302–316
86. Yaglou CP (1949) Indices of comfort. In: Newburgh LH (ed) Physiology of heat regulation and the science of clothing, 1968 edn. Hafner Publishing Co., New York
87. Yi W, Chan APC (2017) Effects of heat stress on construction labor productivity in Hong Kong: a case study of rebar workers. Int J Environ Res Public Health 14:1055
88. Yi W, Chan APC, Wang X, Wang J (2016) Development of an early-warning system for site work in hot and humid environments: a case study. Autom Constr 62:101–113
89. Yi W, Zhao Y, Chan APC, Lam EWM (2017) Optimal cooling intervention for construction workers in a hot and humid environment. Build Environ 118:91–100
90. Yi W, Chan AP (2015) Which environmental indicator is better able to predict the effects of heat stress on construction workers? J Manage Eng 31
91. Zander KK, Botzen WJ, Oppermann E, Kjellstrom T, Garnett ST (2015) Heat stress causes substantial labour productivity loss in Australia. Nat Clim Change 5:647–651

Dr. Ruwini Edirisinghe, B.E., Ph.D. is a vice chancellor's research fellow with expertise in digital innovation (BIM), information and communication technologies (ICT) and smart technologies in building and construction field.

Dr. Mary Myla Andamon, B.Sc. Arch., Ph.D. is a lecturer and a building scientist with expertise in indoor and outdoor thermal comfort studies, urban microclimate and building environmental monitoring and performance assessments.

The Built Environment and Energy Efficiency in Australia: Current State of Play and Where to Next

Trivess Moore and Sarah Holdsworth

Abstract This chapter provides a review and critique of the development and current status of approaches to improve energy efficiency and broader sustainability in the Australian built environment. The focus is on the minimum building performance requirements set through the National Construction Code—Building Code of Australia, but the chapter also includes other mandatory and voluntary approaches which have been introduced over the past two decades. The chapter concludes with a discussion that highlights current gaps that relate to the delivery of a low-carbon/low-energy built environment. It recognises that Australia currently fails to meet international building performance best practice standards, particularly in the residential sector, and this situation can only be reversed if various levels of government in Australia increase the regulated level of energy performance of buildings coupled with a more holistic and progressive inclusion of all energy consumed and generated within a building. This would be better aligned with improving actual impacts a building has over its lifecycle and on the community at large.

1 Introduction

As discussed in chapter one, the built environment has a significant role to play in a transition to a low-energy and low-carbon future. Federal, state and local levels of government in Australia have addressed energy and sustainability performance of the built environment over recent decades. The key policy focus has been through a range of regulatory approaches which includes the introduction of minimum energy performance standards for buildings and appliances and the provision of market-based mechanisms such as rebates for renewable energy technologies [8,

T. Moore (✉) · S. Holdsworth
Sustainable Building Innovation Laboratory, School of Property Construction and Project Management, RMIT University, Melbourne, VIC, Australia
e-mail: trivess.moore@rmit.edu.au

S. Holdsworth
e-mail: sarah.holdsworth@rmit.edu.au

P. Rajagopalan et al. (eds.), *Energy Performance in the Australian Built Environment*, Green Energy and Technology, https://doi.org/10.1007/978-981-10-7880-4_4

18]. This chapter focuses on the current suite of mandatory and voluntary building performance standards and tools, rather than market-based mechanisms which are explored elsewhere [18, 26].

To provide context, building and performance standards are not addressed in the Australian constitution but are set nationally through the National Construction Code (NCC) series, which comprises of the Building Code of Australia (Volumes 1 and 2) and Plumbing Code of Australia (Volume 3) [8]. The NCC is developed and maintained by the Australian Building Codes Board (ABCB). The ABCB represents all levels of Australian Government and has the responsibility of maintaining and developing, through consultation with representatives of the building industry and broader community, the regulatory environment to protect health, safety, sustainability and amenity of people in their use of buildings. Through the NCC, the ABCB ensures the achievement of a consistent set of national building standards, specifically the 'minimum necessary requirements' for matters such as structure, fire resistance, access and egress, services and equipment, energy performance and indoor conditions [2, 3]. The ABCB is the sole authority that provides provision for the design and construction standards of buildings and other structures and in conjunction with the NCC establishes uniform practice, within the industry, to safeguard residents and occupants of buildings as well as protecting building industry stakeholders [2, 3].

The NCC has its origins in Australia as far back as the 1960s [8, 25]. Prior to this time, individual states and territories were left to deal with housing performance regulations, if they did at all. The first attempt at a national code was released in 1971, with the release of the Australian Model Uniform Building Code. This initial code was based on the Local Government Act of NSW, and many states found that they were required to alter the code to fit the particular requirements of their own state and had difficulty with the usability of the format. It was not until 1990 that the first 'useable' national technical building code was developed, known at Building Code of Australia 90. A major revision to the code in 1996 saw the performance-based building code moves away from the prescriptive-based building code. In 2011, the Building Code of Australia and Plumbing Code, which until that point had been separate codes, were integrated under the National Construction Code [8].

The NCC, despite being a national code, due to its historical evolution and vastly different environmental conditions across Australia, must adhere to the building act of each Australian state or territory. As an example, the Victorian Building Act of 1993 governs building activity in Victoria and it sets out the legislative framework for the regulation of building construction, building standards and the maintenance of specific building safety features [42]. The Act has many objectives one of which includes the protection of health and safety of those who use buildings and places of public entertainment. It also seeks to ensure that building amenity such as natural light, air ventilation and thermal comfort of a building are improved [42].

Until 2015, the NCC was reviewed on an annual basis by the ABCB (published annually in May). This resulted in changes to the NCC that were reactive in nature as they address issues with the existing NCC, rather than addressing performance (and other) standards in a longer-term and strategic way. This is further conflated by the fact that the NCC is reviewed based on the code striving to meet minimal performance standards. This regulates for the lowest level of performance the market is prepared to accept; i.e., it places a minimum performance standard a building must comply with [2, 3]. However, recent changes have seen the review period altered to a three-year cycle to provide for an increased compliance consistency to the construction industry. The current version is NCC 2016, with the next update anticipated in 2019.

The former prescriptive approach of the NCC was based on what a building was required to do and presented an acceptable solution to minimum standards [19]. A comparison between the prescriptive and performance-based regulatory approaches can be illustrated using fire safety in a building; a prescriptive approach would state the materials required within the code in order to satisfy fire safety, i.e. what the structural frame should and should not be made of, whereas the performance-based code specifies what the building's structure needs to be, to be able to withstand/be able to perform in line with the objectives of the Act, i.e. how long it needs to maintain its structural integrity during a fire for occupants to escape safely.

The change from the prescriptive approach to a performance approach was a consequence of the recognition that the prescriptive approach failed to promote innovation, improvement, cost optimisation and acted as a barrier to international exchange due to its stringent guidelines and lack of flexibility [14]. In consideration of building production, the flexibility of the performance-based approach enables greater innovation and more economical, higher performing buildings in all aspects of the building code.

Since the introduction of the performance-based Building Code of Australia, there have been a number of reforms, particularly in 2003, 2006 and 2010 [8]. While each round of reforms has seen improvements to some elements of sustainability (e.g. heating and cooling energy requirements), there has been limited requirement for more holistic sustainability improvements and inclusions. Rather, these reforms have focused on issues such as limitations on liability for building practitioners, privatisation of building approvals and inspections, compulsory registration of building practitioners and compulsory insurance for building practitioner. It is important to consider that identified building standards still only prescribe/regulate the minimum requirements of safety health and amenity that a building must manifest for its users [2, 3].

As discussed in chapter one, the built environment is responsible for significant greenhouse gas emissions and consumption of energy [22, 35, 36]. In this context, energy regulation has been a key focus area in the policy and regulatory landscape for the construction industry. This has been driven locally by international agreements and accords such as the Kyoto Protocol which lay the responsibility of carbon emission reduction at the national and, in turn, local levels. Australia has addressed this challenge through the adoption of energy efficiency techniques and new technologies throughout the construction industry. How this has evolved within the Australian construction industry, both residential and non-residential sectors, will be explored below. This will precede a discussion on where this leaves energy (and sustainability) performance for the built environment in Australia and its future direction.

2 Residential Sector: Energy Performance and Regulation

In Australia, minimum energy performance requirements are set through the NCC as guided by the Council of Australian Governments (COAG) and the ABCB under the Building Act. The current energy performance requirements of the code are reiterated through the Nationwide House Energy Rating Scheme (NatHERS). NatHERS was initially developed in 1993 as a structure to rate or evaluate the energy performance of dwellings [21]. NatHERS provides guidelines for how to rate a dwelling, with accredited third-party software, such as AccuRate, FirstRate5 and BERS Pro, doing the actual computer simulation of the building's energy performance.

The rating system was developed as a tool to calculate energy and thermal performance for standard and more sustainable housing designs. The tool's focus and goals are to improve the energy performance of a building through its design and construction. Initially, NatHERS looked primarily at heating and cooling energy requirements but has been expanded, in the 2010 revision, and in the subsequent decade to include some requirements for fixed lighting and hot water [8]. NatHERS results are ranked across a 'star' rating band. The star band ranges from 1 star (least natural thermal performance) to 10 star (best natural thermal performance, requiring virtually no mechanical heating and cooling) [31]. The required performance within each star band is adjusted based upon climatic conditions within a specific location. Within NatHERS, there are 69 different climate zones. Figure 1 presents the star bands for the capital cities of Australia, demonstrating the different requirements to achieve each star band. The NatHERS rating system currently requires that all new and renovated residential buildings meet a minimum 6-star energy rating.

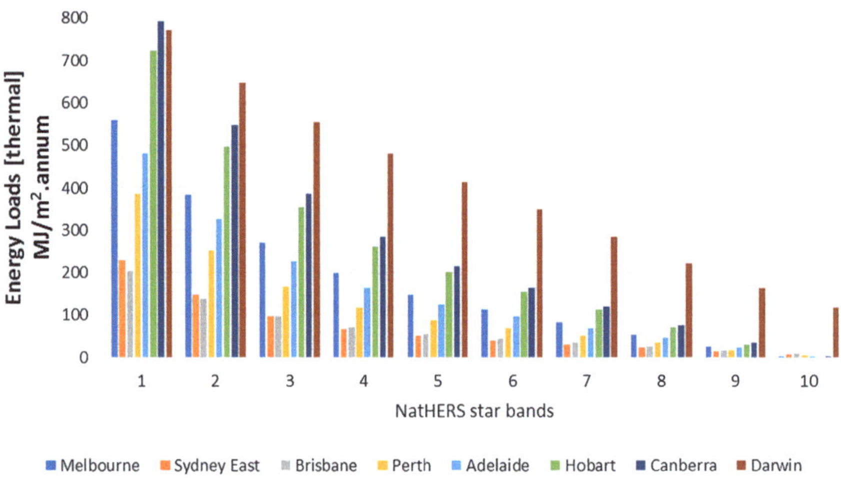

Fig. 1 Thermal energy load requirement for the different star bands for the capital cities in Australia [31]

The introduction of NatHERS and minimum energy performance requirements within the NCC has had significant impact on the improvement of energy performance across new and renovated building stock. Pre-1990, when policy requirements for minimum housing insulation were first introduced, existing housing had an average performance standard of 1 star. This rose to an average of 2.2 stars after the introduction of minimum insulation standards [31]. A minimum star rating requirement for new housing was not set until 1997, when the Australian Capital Territory introduced a requirement that all new housing be built to a 4-star NatHERS minimum energy performance standard [13]. Victoria was the next state to introduce mandatory minimum star rating performance [8]. From 2004, Victoria set 5-star NatHERS as the minimum performance standard for new housing. The 2004 standard included an extra requirement for either onsite solar hot water or a rainwater tank to be incorporated on all new residential buildings. Other states and territories followed over the next few years until the current 6-star NatHERS minimum was introduced and adopted by all states in 2011.

While the move to 6-star NatHERS minimum energy performance is on the one hand a significant step forward, it is a standard which still falls short of what is required for a transition to a low-energy/carbon future [6, 25, 46]. This is both because the requirements have a focus on improving thermal efficiency and therefore do little to address other energy consumption such as appliance, and also because a 6-star NatHERS design still requires significant energy for heating and cooling. It is not until a house is nearer to a 9-star standard that the requirement for mechanical heating and cooling is eliminated in line with requirements for a low-energy house (see Fig. 1).

There has been an ongoing challenge for any meaningful public discourse about the next national improvement step for NatHERS. The recent change for the NCC from yearly to three yearly revision cycles is likely to delay any improvements even further. While some states are setting more stringent greenhouse gas emission reduction targets, higher renewable energy and energy efficiency targets, this has not yet translated to an increase in minimum energy performance represented by the minimum NatHERS requirements. While COAG, ideally, desires a national consensus aligned to improved future energy efficiency policy, regulations and outcomes, the states and territories have the ability to legally set their own requirements beyond those set in the NCC [8].

In the absence of an increase in the energy efficiency requirements to the minimum NatHERS regulations, local government planners, through amendments to state-based planning schemes (e.g. the Victorian Planning Scheme), have increased performance regulation requirements in an attempt to drive improved sustainability outcomes in the built environment. For example, BASIX (Building Sustainability Index) is an additional sustainable housing measure which sets energy and water performance requirements in excess of the 6-star NatHERS minimum requirements. BASIX is a tool which must be used during the development application process in the state of New South Wales to help achieve planning permission. Introduced in 2004, BASIX has recently been revised to improve its performance standards in line with broader goals of energy and water efficiency across the residential sector in New South Wales. Other examples of planning requirements in New South Wales include the State Environmental Planning Policy No 65—Design Quality of Residential Apartment Development (SEPP 65).

Within Victoria, similar planning initiatives include the Apartment Design Guidelines which require improved thermal and energy performance of apartments and more general design principles such as minimum room sizes and access to natural light [12]. Additionally, a group of local councils in Victoria also have the requirement to use the Building Environment Sustainability Scorecard (BESS) as part of planning applications in their jurisdictions to improve sustainability outcomes [28]. Further, while not statutory in all councils, the Ministerial approval and formal gazettal of Amendment C133 introduces Clause 22.17 *Environmentally Sustainable Development* to some Planning Scheme. Six leading councils in Melbourne developed the Planning Scheme amendment via a cross-council collaboration with the cities of Banyule, Moreland, Port Phillip, Stonnington, Yarra and Whitehorse and have adopted the clause in their Planning Scheme [28]. The overarching objective is that development should achieve best practice in environmentally sustainable development from the design stage through to construction and operation. The guidelines apply to single dwellings, apartments, town houses, commercial buildings and warehouse conversions. They include examples of building design and layouts that optimise natural daylight, cross-ventilation and sunlight all year round and provide guidance on the selection of energy-efficient building materials and appliances.

While the above approaches are requirements at the time of design and construction, there have also been attempts at market-based information provision to drive improvements in residential energy performance in Australia. For example, the Australian Capital Territory introduced mandatory disclosure in 1999. Initially required under the House Energy Rating Scheme, it is now administered under the *Civil Law (Sale of Residential Property) Act 2003*. The *Act* requires the disclosure of an existing dwelling's energy rating in all advertisements for sale/rental of a residential building and that the contract of sale includes information about the building's Energy Efficiency Rating. Research on this scheme found that there was an increased resale value for higher-star-rated dwellings [13]. While other states have discussed introducing similar mandatory disclosure requirements, this has not eventuated. However, the Victorian Government rolled out a Residential Energy Scorecard in 2017 which is a voluntary tool which they hope consumers will use for providing information at the point of sale/lease of their properties. At this stage, there is no plan to make that tool mandatory.

3 Non-residential

The minimum energy performance (and broader sustainability elements) standards for non-residential buildings are set within the NCC, as per the residential sector, specifically, through Section J of the code. One of the key objectives of Section J, in the BCA, is to reduce greenhouse gas emissions. Mandatory compliance performance-based BCA was a government intervention due to market failure in the development of improved energy performance. Within Section J, Australia is divided into eight climate zones and these are used to determine the thermal design of the building.

In the building code, there exists regulation on the proposal of new or alternate products and how these can meet performance requirements. Alternate products are required to pass performance requirements through a verification method. This verification method allows for innovation and promotes the better use of a building's fabric and services to make a building more energy-efficient [1]. Before an alternate product can be used, a theoretical annual energy consumption of the proposed product is calculated in comparison with a base reference product. If the alternate product's score is deemed greater than the reference product, then the alternate product meets the BCA's performance criteria [2, 3]. This lag in time can be problematic with the adoption of new technologies and innovations. Due to each innovation having to be reviewed through a lengthy and rigorous testing process, these new/alternate products may be discarded in favour of known products that have already passed the performance criteria. So, instead of the incentive to utilise/develop higher performing products and building practices, the current system may de-incentivise innovation due to a lag in implementation and market adoption.

While the NCC sets minimum standards of performance, Green Star is a voluntary tool which targets innovation in high performing buildings, like LEED in the USA and BREEAM in the UK. Green Star was developed by the Green Building Council of Australia (GBCA). The first tools under Green Star were launched in 2003. GBCA developed a range of different Green Star tools which addressed different building types before streamlining the tool range to just four from 2016. The tools available now are designed to cover all non-residential building types (and some residential) and include:

- Design and as-built,
- Interiors,
- Performance and
- Communities.

As with other rating tools around the world, the focus of Green Star has shifted from a design tool to a design and as-built rating and looks to encourage the ongoing verification of performance through the Performance tool. This is to try and reduce gaps between design intent and actual performance outcomes. As of November 2017, there are more than 1700 certified Green Star buildings in Australia, a number that is growing each year (Fig. 2). Evaluations from Green Star certified buildings have found significant benefits including consuming 51% less potable water, consuming 66% less electricity, producing 62% less emissions, recycling 96% of construction waste and a range of other health, wellbeing and productivity improvements for occupants in the buildings, all achieved within a 3% capital cost premium [15, 16].

The Green Star rating certification scheme has resulted in a greater focus in the actual performance verification; however, this assessment is still voluntary and has

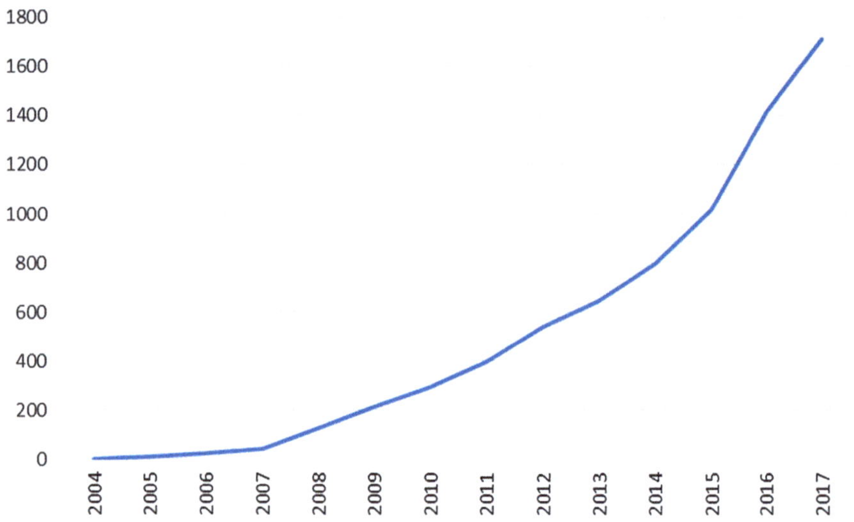

Fig. 2 Number of Green Star accredited buildings from 2004 to 2017 [15, 16]

not resulted in a whole of sector transformation. Mandatory commercial energy performance and disclosure exists in Australia but only in the area of energy consumption during the operation of a building. The National Australian Built Environment Rating System (NABERS) is a tool which evaluates actual performance via four tools for Energy, Water, Indoor Environment Quality and Waste Generation across a scale of 1 (worst)—6 (best) stars [33, 34]. Offices with a floor area greater than 1000 m^2 are required to disclose their energy performance with a NABERS rating when the space is being leased or sold. It is a voluntary tool for spaces below this size, or for other building types (e.g. NABERS have a tool which can be applied for housing). A rating is given based upon the previous 12 months' worth of data, and the rating itself is only valid for the next 12 months. NABERS is now also able to be used to validate performance within Green Star.

To date, NABERS Energy has now been used to rate more than 72% of Australian office space [33, 34]. Office buildings using NABERS have reported an average improvement in energy efficiency of almost 10%. Similar savings have been found for office buildings using NABERS Water (11% saving). Research has also identified that 5-star NABERS-Energy-rated buildings have been found to deliver a 9% green premium in value and the 3–4.5-star NABERS Energy ratings deliver a 2–3% green premium in value [33, 34].

Building upon NABERS, the federal government introduced the Commercial Building Disclosure Program which requires sellers and lessors of office space of 1000 m^2 or more to have an up-to-date Building Energy Efficiency Certificate (BEEC) [5]. A BEEC is necessary to comply with legal obligations under the Building Energy Efficiency Disclosure Act 2010 (revised to Building Energy Efficiency Disclosure Amendment Act 2015). The BEEC is an assessment of energy performance of a building or area of a building but does not consider green energy in its assessment. The assessment contains three parts:

- Part 1 consists of a NABERS Energy for offices rating.
- Part 2 consists of a tenancy lighting assessment.
- Part 3 is a guidance list for the building/area.

The BEECs are valid for up to 12 months for Part 1; a Part 2 evaluation is valid for up to 5 years.

4 Discussion

The urgency to improve energy performance was solidified when 180 nations, including Australia, signed the Paris Agreement in 2016, to commit to actions that would keep a global temperature rise below 2 °C above pre-industrial levels this century [41]. This committed Australia to tangibly addressing climate change and move towards a reality of a 'net zero' economy by 2050. As discussed in the preceding sections of this chapter, like many countries, over the past three decades, Australia has developed a set of minimum energy performance requirements for new and renovated buildings [8, 26]. Alongside this there has been the development of voluntary rating tools, design guidelines, planning regulations and other market mechanisms (e.g. solar photovoltaic rebate schemes) which have helped to drive the top end of the performance market. The Australian Sustainable Built Environment Council released a report in 2016, *Low Carbon, High Performance,* which claimed that Australia can eliminate emissions from the built environment entirely by 2050 [4]. Other groups have also found that similar outcomes are feasible for Australia [9].

The impact buildings have on the environment and the community which live and work within them requires a greater understanding and application of performance standards in the construction and associated use of buildings. However, current minimum requirements in Australia lag behind those of international best practice and are significantly lower than performance outcomes required for a transition to a low-carbon and energy future [25]. Under Australia's current target, Australia's per capita emissions would be around three times higher than global per capital emissions [40]. Clearly, the introduction of NatHERS, Green Star, NABERS and other policy and regulatory drivers has resulted in documented financial, environmental and social improvements; however, more action is required if the desired low-energy and carbon future is to be achieved.

Within professions of the sustainability sector and across the broader community, there has been a push to tighten energy efficiency within the NCC. However, despite the recognition of the ABCB and COAG that low-energy/carbon housing is quickly becoming international best practice, both governing organisations have had limited participation in this emerging discourse and development of policy in this area [11]. There is probably no clearer example of the fact that 'Australia has entered a building energy policy hiatus' [8: 964] than the fact that there has been no systemic review of domestic energy efficiency for the past 10 years. This last review led to the introduction of the minimum 6-star NatHERS standard in 2011. Given the new approach to the review of the building code every three years, there will be a considerable lag between the development of innovative building materials and techniques and the adoption of them into the construction industry or the increase of the minimum requirements themselves.

Two contributing factors to the outdated practices occurring in Australia are the assumptions embedded in the definition of performance which underpins the BCA (the protection of building inhabitants through its structural safety) and the performance approach only requiring minimal performance. As a result of this inertia, most buildings simply meet the status quo and industry stakeholders fail to take the necessary steps to significantly increase the performance of their buildings or adapt to best practices in the case of new buildings. While the non-residential sector is achieving improved building performance through the Green Star initiative, at the top end of the market, and helping innovate around sustainability and energy efficiency, there is no equivalent in the domestic housing sector that is being taken up by the market. In this context, it is important that the minimum performance requirements for housing be improved in the short term, not just by increasing the minimum NatHERS requirement, but also for revised minimum performance requirements to include all energy consumed within a dwelling, and include requirements for onsite renewable energy generation where feasible.

Moore [25] explores what a step change process could look like in Australia over a short–medium term whereby both NatHERS requirements are increased one star at a time from 6–8 stars, but also a stepped increase in requirement for inclusion of renewable energy generation for detached housing in Victoria. This must happen with a clear 10–15-year plan such as achieving net zero energy buildings by 2030, an approach which would provide time for the industry and consumers to adjust, and for industry to innovate and find affordable ways to deliver such housing. In addition, a review of building performance standards against international standards and signatory agreements (such as the UN Development Goals and Carbon mitigation imitative) to curb emission regularly could be built into the NCC review process.

Increased cost of building best practice construction often inhibits the recognised improvements despite the recognised improvements in building performance. Additionally, here is uncertainty about its acceptance within the community with many in the industry already claiming they meet best practice standards. The issue of cost further problematises the issue of performance, and the BCA is that within the Act there exists a contradiction between performance and affordability. When defining and regulating for energy-efficient buildings, the Act regulates for 6-star (NatHERS) energy-efficient housing in the context of affordability. The Act states that it only facilitates cost-effective construction and maintenance of building and plumbing systems. It is opinions like these that continue to present barriers to further progress [38]. Australia's stance to improving this gap can be found in their reluctance to bring in national policy/road maps to improve sustainable development as other jurisdictions such as the EU, UK and California have done [25, 32]. This is further supported by an international study undertaken by Price Waterhouse Cooper [37], which identified key barriers to the adoption of innovation in construction including little information regarding plans and pathways; clear definitions and measuring tools; cost of technology; stakeholder awareness; skilled experts and government incentives. We are still facing many of these key barriers today with limited progress towards addressing them in Australia.

Further adding to the complexity and issue of performance is the standards for the construction and maintenance of buildings contained in the BCA, compliant with the Building Act 1993. The issue is that the building regulations are founded on type not use (Class); they do not allow for the identification of individual buildings, which may have amenity issues associated with their use or the use of surrounding buildings [10]. An example is the development of a medium-density apartment complex next to a live music venue. This is problematic when considering the cumulative impacts of buildings in relation to their environmental and social performance.

5 Where to from Here

If buildings in Australia are to improve their energy and sustainability performance, then a change in policy, standards, buildings produced and community acceptance of the need to improve efficiency will need to occur. Currently, new and renovated buildings are typically being delivered to meet minimum standards rather than striving to achieve superior market standards. Problematic is the fundamental assumption associated with the term performance, as defined in the BCA. Building and construction underpinned by paradigms of sustainability would require a definition of performance that is more encompassing and long term. Buildings would not only ensure their performance related to the physical safety of those within them now but also how their performance contributed to the quality of the environment in the future, i.e. consideration of a building's energy performance across its life cycle, both directly and indirectly. In this context, with a building's life assumed to be 40 years by the ABCB, we need to move from using historical climate data to using climate predictions which at least represent the half way point of the building's life cycle, to ensure that the best performance is achieved across the whole life. Research has shown that the existing performance of housing is likely to worsen in many climate zones around Australia as the natural climate changes [43]—therefore we are handicapping our housing before it is even built by the assumptions included in the sustainability ratings [7, 29]. Further international benchmarking of associated building standards and a proactive approach to NCC revisions would see ongoing performance improvement of building rather than stagnation.

The definition of performance should also include the social, and not simply human, capital of their occupants, and other researchers have identified that there is a requirement to bring the wider health, wellbeing and social benefits into the regulation debate [17, 30, 44, 45, 47] (see also Chapters "Low-Energy Housing as a Means of Improved Social Housing: Benefits, Challenges and Opportunities", "Are We Living With Our Heads in the Clouds? Perceptions of Liveability in the Melbourne High-Rise Apartment Market" and "The Way Forward-Moving Towards Net Zero Energy Standards"). For example, in jurisdictions such as the UK, EU and California, pathways to low-energy housing have been set out which linked these requirements to other government goals such as greenhouse gas emission reduction targets, renewable energy generation targets, improved thermal comfort targets and reducing fuel poverty rates [27]. This likely gives the requirements for improved energy efficiency greater support.

There is also a need for other changes across the building industry if high performing buildings are to be delivered on mass. This includes getting the peak industry associations to support such an outcome. Currently, many of these peak stakeholder organisations are cautious about any changes to minimum energy performance regulations as they fear this will add costs to projects, which will need to be passed onto consumers who are already facing affordability constraints [20, 23]. Increasingly though, research is demonstrating that improved energy and sustainability performance can be delivered for low, or no, additional capital cost and that the building industry is able to adjust quickly to find economic efficiencies [24, 39].

Any move to improve sustainability requirements will also need to be accompanied by improved training of those designing and constructing the buildings. It is vital to ensure that the skills required to achieve alternative (i.e. efficient) products are available to the market. Educating and building increased human capital in these emerging areas of the industry would only be helped by a clear longer-term strategic plan so that the building industry can understand how they need to adapt and prepare [27]. Some of this is occurring in Australia; for example, the Master Builders Association run a Green Living programme to help provide builders with the required tools to engage with improved sustainability in their construction; however, more needs to happen across all stakeholders.

We know that we must be delivering low- or zero-energy/carbon housing and that we have the technologies, materials and design skills to do so. The main challenge as identified by Berry and Marker [8] is that there is a lack of political will in Australia to improve minimum performance requirements for buildings. The longer that Australia takes to develop a pathway to low-energy or carbon housing, the further we fall behind international best practice and the harder it will be for Australia to meet greenhouse gas emission reduction targets, not to mention the longer-term 'lock-in' problems for owners, occupants and broader society, which will be created by adding more unsustainable housing to our current poorly performing stock.

6 Conclusion

A number of approaches have been implemented by various levels of government in Australia over the past few decades to address the energy and sustainability performance of the built environment. This chapter has explored the development of mandatory regulations (i.e. building codes) and voluntary approaches in the Australian context. While the current provision of buildings in Australia has improved energy and sustainability performance in comparison with previous development, and minimum performance requirements still fall significantly short of what is required for a transition to a low-/zero-carbon and energy future. Australia is lagging behind international best practice in terms of building performance and has not articulated a public position for future performance requirements (see Chapter "The Way Forward-Moving Towards Net Zero Energy Standards"). There are a number of opportunities for policy improvement which would help guide the building industry, and consumers, towards improved performance of buildings in Australia, with recent research highlighting benefits for occupants (see Chapter "Low-Energy Housing as a Means of Improved Social Housing: Benefits, Challenges and Opportunities") but also the need to understand occupants and changing housing needs in this push towards a more sustainable built environment (see Chapter "Are We Living With Our Heads in the Clouds? Perceptions of Liveability in the Melbourne High-Rise Apartment Market"). However, the lack of political will to address this is a significant barrier which shows no signs of being overcome in the near term.

References

1. ABCB (2016) NCC volume two energy efficiency provisions 2016, 3rd edn. Australian Building Codes Board, Canberra
2. ABCB (2017a) About. Retrieved 12 Aug 2017, from http://www.abcb.gov.au/NCC/About
3. ABCB (2017b) Code compliance. Retrieved 12 Aug 2017, from http://www.abcb.gov.au/NCC/How-it-works/Code-Compliance
4. ASBEC (2016) Low carbon, high performance. How buildings can make a major contribution to Australia's emissions and productivity goals. Australian Sustainable Built Environment Council, Sydney
5. Australian Government (2017) What is a BEEC? Retrieved 04 Sep 2017, from http://www.cbd.gov.au/get-and-use-a-rating/what-is-a-beec
6. Berry S (2014) The technical and economic feasibility of applying a net zero carbon standard for new housing. University of South Australia
7. Berry S, Davidson K (2016) Improving the economics of building energy code change: a review of the inputs and assumptions of economic models. Renew Sustain Energy Rev 58(Supplement C):157–166
8. Berry S, Marker T (2015) Residential energy efficiency standards in Australia: where to next? Energ Effi 8(5):963–974
9. BZE (2013) Zero carbon Australia: buildings plan. Beyond Zero Emissions, Melbourne Energy Institute, University of Melbourne, Melbourne
10. Carbines E (2003) Live music taskforce: report and recommendations, December 2003. Department of Sustainability and Environment, Victorian Government, Melbourne

11. COAG (2009) National strategy on energy efficiency. Canberra, Council of Australian Governments
12. DELWP (2017) Apartment design guidelines for Victoria. Department of Environment, Land, Water and Planning, Melbourne
13. DEWHA (2008) Energy efficiency rating and house price in the ACT. Department of the Environment Water Heritage and the Arts, Canberra
14. Foliente GC (2000) Developments in performance-based building codes and standards. For Prod J 50(7/8):12–21
15. GBCA (2017a) Green Star. Retrieved 04 Sep 2017, from http://new.gbca.org.au/green-star/
16. GBCA (2017b) Green star project directory. Retrieved 04 Sep 2017, from http://www.gbca.org.au/project-directory.asp
17. Golubchikov O, Deda P (2012) Governance, technology, and equity: an integrated policy framework for energy efficient housing. Energy Policy 41:733–741
18. Greene D, Pears A (2003) Policy options for energy efficiency in Australia. Melbourne, ACRE and The Australian CRC for Renewable Energy Policy Group
19. Gross JG (1996) Developments in the application of the performance concept in building. In: 3rd CIB-ASTM-ISO-RILEM international symposium, Haifa, Israel, National Building Research Institute
20. HIA (2009) Response to the consultation regulation impact statement 2009-03. Proposal to revise the energy efficiency requirements of the BCA for residential buildings. Housing Industry Association, Campbell, ACT
21. Kordjamshidi M (2011) House rating schemes from energy to comfort base. Springer, Berlin
22. MA (2005) Ecosystem and human well-being: opportunities and challenges for business and industry—the fourth Millennium Ecosystem Assessment report. Millennium Ecosystem Assessment
23. MBAV (2008) Submission to the liveability inquiry. Victorian competition and efficiency commission, February 2008. Master Builders Association of Victoria, Melbourne
24. MEFL (2017) Changes associated with efficient dwellings project—Final report Melbourne, Prepared by Moreland energy foundation in association with strategic, policy and research. WTP partnership and building environmental assessment company on behalf of the department of the environment and energy
25. Moore T (2012) Facilitating a transition to zero emission new housing in Australia: costs, benefits and direction for policy. RMIT University
26. Moore T, de Haan F, Horne R, Gleeson B (2018) Urban sustainability transitions. Australian cases—international perspectives. Springer, Singapore
27. Moore T, Horne R, Morrissey J (2014) Zero emission housing: policy development in Australia and comparisons with the EU, UK, USA and California. Environ Innov Societal Transitions 11:25–45
28. Moore T, Moloney S, Hurley J, Doyon A (2017) Implementing sustainability in the built environment. An analysis of the role and effectiveness of the building and planning system in delivering sustainable cities. RMIT University, Melbourne
29. Moore T, Morrissey J (2014) Lifecycle costing sensitivities for zero energy housing in Melbourne, Australia. Energy Build 79:1–11
30. Moore T, Strengers Y, Maller C (2016) Utilising mixed methods research to inform low-carbon social housing performance policy. Urban Policy and Res 34(3):240–255
31. NatHERS (2017) Nationwide House Energy Rating Scheme (NatHERS). Retrieved 12 Aug 2017, from http://www.nathers.gov.au/
32. Newton P, Newman P (2013) Low carbon green growth: tracking progress in Australia's built environment industry towards a green economy. Swinburne University of Technology and Curtin University, Melbourne
33. NSW Government (2017) An introduction to NABERS. Retrieved 04 Sep 2017, from https://nabers.gov.au/public/webpages/ContentStandard.aspx?module=10&template=3&include=Intro.htm&side=EventTertiary.htm

34. NSW Government (2017) Research and statistics. Retrieved 04 Sep 2017, from https://nabers.gov.au/public/webpages/ContentStandard.aspx?module=10&template=3&include=ResearchStats.htm&side=factsheets.htm
35. OECD (2003) Environmentally sustainable buildings: challenges and policies. OECD Publishing, Paris
36. Porritt J (2007) Capitalism as if the World matters. Earthscan, London
37. Price Waterhouse Cooper (2002) Innovation in the Australian building and construction industry. Report prepared for Australian building and construction industry. Forum and the Department of Industry, Tourism and Resources
38. Reed R (2008) The relationship between sustainability and housing value. In: 8th annual African real estate society conference. Johannesburg, South Africa
39. Sustainability House (2012) Identifying cost savings through building redesign for achieving residential building energy efficiency standards. Prepared for the Department of Climate Change & Energy Efficiency, Canberra
40. The Climate Institute (2015) Factsheet: Australia's emissions: what do the numbers really mean? The Climate Institute
41. United Nations (2015) Framework convention on climate change. Adoption of the Paris agreement, 21st conference of the parties. United Nations, Paris
42. VBA (2017) Legislation. Retrieved 04 Sep 2017, from http://www.vba.vic.gov.au/practitioners/legislation
43. Wang X, Chen D, Ren Z (2010) Assessment of climate change impact on residential building heating and cooling energy requirement in Australia. Build Environ 45(7):1663–1682
44. Wells W, Bardacke T, Cepe P, Seal Cramer J, McManigal Delaney L, Landman M (2007) Blueprint for greening affordable housing. Island Press, Washington
45. Willand N, Ridley I, Maller C (2015) Towards explaining the health impacts of residential energy efficiency interventions—a realist review. part 1: pathways. Soc Sci Med 133:191–201
46. Williams J, Mitchell R, Raicic V, Vellei M, Mustard G, Wismayer A, Yin X, Davey S, Shakil M, Yang Y, Parkin A, Coley D (2016) Less is more: a review of low energy standards and the urgent need for an international universal zero energy standard. J Build Eng 6(Supplement C):65–74
47. Williamson T, Grant E, Hansen A, Pisaniello D, Andamon M (2009) An investigation of potential health benefits from increasing energy efficiency stringency requirements. Building code of Australia volumes one and two. The University of Adelaide, Adelaide

Dr. Trivess Moore B.Sc. (Env.), Ph.D. is a research fellow with expertise in energy efficiency, renewable energies, the built environment and transitioning to a low-carbon future.

Dr. Sarah Holdsworth B.Ed. (Env. Sci.), M.Env.Sci., Ph.D. is Senior Lecturer in sustainability with expertise in sustainable education for the building industry, specifically notions of sustainability in relation to curriculum development and professional education.

Environmental Rating Systems for Non-Residential Buildings—How Does Australia Compare with International Best Practice?

Priyadarsini Rajagopalan

Abstract Growing concerns over negative impacts associated with buildings have compelled governments across the world to introduce minimum requirements for energy efficiency. Energy and environmental performance rating tools and minimum energy performance standards have become widespread in the last two decades. This chapter reviews the status of environmental rating systems in the non-residential building sector in Australia and compares with other leading international rating systems with a focus on those relating to new building design and construction. The major non-residential rating system in Australia, Green Star, was introduced in 2003 by the Green Building Council of Australia and is broadly comparable to international tools such as LEED and BREEAM. While Green Star has been an important driver of improving energy efficiency in non-residential buildings in Australia, it has suffered from inconsistent commitment to climate action from both major Australian political parties. Even though Green Star has similar criteria and performance standards in comparison to LEED and BREEAM, the market penetration of this rating system falls behind other systems in terms of adoption rate. Proper government support and improvement of supply chains would certainly help the rating systems to penetrate the wider market.

1 Introduction

As the impacts of climate change have become more obvious, there is worldwide interest in saving the environment and natural resources. The built environment is one of the largest contributors to greenhouse gases as discussed in Chapter "The built environment in Australia". Many countries have been making substantial efforts to reduce the impacts of climate change by adopting various mitigation strategies such as mandatory and voluntary energy labelling schemes [5, 12]. While local-level

P. Rajagopalan (✉)
Sustainable Building Innovation Laboratory, School of Property Construction and Project Management, RMIT University, Melbourne, VIC, Australia
e-mail: priyadarsini.rajagopalan@rmit.edu.au

P. Rajagopalan et al. (eds.), *Energy Performance in the Australian Built Environment*, Green Energy and Technology, https://doi.org/10.1007/978-981-10-7880-4_5

mitigation efforts in Australia emerged during the 1990s and 2000s, it was not until the late 2000s that a national response to climate change was articulated [26]. Climate change is a long-term problem that requires stable but flexible policy implementation over time. However, Australia's commitment to climate action over the past three decades has been inconsistent and lacking in direction [36]. The climate policies of the two major Australian political parties have varied considerably over the years.

As introduced in Chapter "The built environment and energy efficiency in Australia: current state of play and where to next", the Green Building Council of Australia (GBCA) was established in 2002 to lead the sustainable transformation of Australia's built environment. In 2003, GBCA launched the Green Star environmental rating system which has become the largest voluntary rating tool in Australia. Despite the size of the Green Star programme and the trend in the market towards greater energy efficiency, independent research exploring market adoption is limited [33]. Most of the self-reported literature published by GBCA comparing the certified floor–area serves as marketing material. For example, one recent report [14] shows the total number of buildings certified using the system but does not compare this with the total number of commercial buildings built during the time period or the total number of building stock, making results of limited value.

This chapter builds upon the introduction of non-residential energy efficiency approaches in Australia presented in Chapter "The built environment and energy efficiency in Australia: current state of play and where to next". The chapter reviews the environmental rating systems in the non-residential building sector in detail, with particular emphasis on the Green Star rating system for new buildings and its uptake within the industry, and locates this within a discussion of international best practice.

2 Energy Efficiency Policies in the Australian Built Environment

Various policies and incentives have been implemented in Australia to promote the adoption of green buildings. Table 1 gives an overview of recent policies in relation to energy and environmental performance of non-residential buildings (see Chapter "The built environment and energy efficiency in Australia: current state of play and where to next" for further details). Following changes to the National Construction Code and the Australian government's ratification of the Kyoto Protocol, a number of other approaches have been introduced. In 2009, the Green Building Fund was launched to provide A$90 million as financial incentives to assist building owners to improve the energy efficiency of their buildings. In 2010, the Commercial Building Disclosure policy came into place because of the Building Energy Efficiency Disclosure Act. The Commercial Building Disclosure scheme ensures that when a commercial building with a net lettable area of 2000 m^2 or more is sold or leased, the energy efficiency information of the building must be provided. By making it mandatory to disclose this information, it ensures that the renters or buyers can make

Table 1 Australian policies in relation to the commercial building sector

Policies	Year introduced
National Construction Code Energy Efficiency Requirements	2006 (updated 2012)
Energy Efficiency in Government Operations	2006
Ratification of Kyoto Protocol	2008
Green Building Fund	2009
Commercial Building Disclosure	2010
Introduction of Carbon Tax	2011
Abbot Government Repeal Carbon Tax	2013
Environmental Upgrade Agreements	2012 (NSW) 2013 (SA) 2015 (VIC)

an informed decision on the building with respect to the impact it has on the environment. In June 2016, there was a move to lower the mandatory disclosure threshold on commercial office buildings from 2000–1000 m^2 which expected to see an additional 1000 commercial buildings disclose their energy efficiency when they sell or lease their property [24]. It is expected that this will deliver more than A$50 million in energy savings, and around 3.5 million tonnes of emission reductions over five years.

In 2011, the federal Labour government introduced a carbon pricing or "carbon tax" through the Clean Energy Act 2011 to reduce greenhouse gas emissions. The initiative was also intended to support economic growth through the development of clean energy technologies. At the time of implementation, the carbon price only applied to the top 500 carbon polluters in the country. The carbon tax generated intense political debate and faced significant challenges from the public and the federal opposition government. To compensate for the potential increase in fuel price, the governing federal Labour government funded a range of initiatives including energy efficiency measures and renewable energy target and provided direct financial rebates to most households to offset additional costs resulting from the tax. A change of federal government in 2014 resulted in the Liberal government repealing the carbon tax. Thus, Australia became world's first developed nation to cancel carbon laws that put a price on greenhouse gas emissions.

3 International Development of Energy and Environmental Rating Systems

The last 20 years have seen an emergence of built environment energy and environmental programmes in different parts of the world including North America, European Union, South America and Asia [30]. A number of rating tools have been developed in these jurisdictions to rate buildings for environmental credentials with different criteria that can be used at all phases of development including design, construction and operations. Points awarded for each category are generally weighted to calculate an overall score for sustainability. Developed in 1990, the Building Research Establishment Environmental Assessment Method (BREEAM) in the UK was the first tool adopted by the building industry. This was followed by the Leadership in Energy and Environmental Design (LEED) rating system developed by the Green Building Council of the USA in 1998. Subsequently, other rating systems have been developed in other parts of the world and their progress has been reviewed over the years. Janda [19] identified the worldwide status of energy standards for buildings with more focus on developed countries. A survey of 81 countries revealed that 61 countries had some form of mandatory and voluntary existing standards, 11 countries had proposed standards, and 9 countries did not have standards [19]. Bernadi et al. [4] carried out a survey of more than 70 schemes and selected six most studied and adopted schemes including BREEAM and LEED for in-depth analysis. The authors note that a systematic comparison of the schemes is difficult, sometimes even prohibitive. In a study comparing the issues and metrics of five representative assessment schemes around the world including BREEAM, LEED, Comprehensive Assessment System for Built Environment Efficiency (CASBEE), Building Environmental Assessment Method (BEAM) Plus and the Chinese scheme ESGB, Lee [21] states that BREEAM and LEED are the most comprehensive tools with widest scope and many other rating schemes are developed based on these two tools.

The European Union has been leading in the building energy efficiency agenda over the last 15 years. The first version of the Energy Performance Building Directive (EPBD), 2002/91/EC, was approved on 16 December 2002 and entered into force on the 4 January 2003, setting a series of energy performance requirements for existing and new buildings. The main aspects include establishment of a calculation methodology, minimum energy performance requirements, an Energy Performance Certificate and inspections of boilers and air conditioning. In the updated Directive 2010/31/EU, all new buildings shall be nearly zero energy buildings by the 31 December 2020; the same applies to all new public buildings after the 31 December 2018 [13]. BREEAM is the UK's environmentally sustainable certification tool similar to Green Star in Australia. The scheme is composed of ten categories, a percentage-weighting factor is assigned to each category, and the overall number of 112 available credits is proportionally assigned. However, a minimum achievement is required for the categories Energy and CO_2 emissions and Water and Waste. The rating scales that can be achieved by buildings are: outstanding ($\geq$85% points achieved in assessment), excellent ($\geq$70), very good ($\geq$55), good ($\geq$45) and pass ($\geq$30).

In the USA, the Energy Star rating system, jointly operated by the Environmental Protection Agency (EPA), and the US Department of Energy (DOE) is equivalent to the National Australian Built Environment Rating System (NABERS—see Chapter "The built environment and energy efficiency in Australia: current state of play and where to next") and includes only energy and indoor environmental quality as the criteria. The LEED system is a voluntary rating system similar to the Green

Star system and takes a broader approach to assess the environmental character of a building in comparison to Energy Star. Different schemes are designed for rating new and existing buildings. Each scheme has the same list of performance requirements set out in five categories, but the number of credits, prerequisites and available points vary considerably according to the building type. Depending on the credits accrued in each category, certification can range from platinum, gold, silver and the simple achievement of certification.

In Australia, Green Star is a voluntary rating system that assesses the environmental performance of projects at all stages of the built environment life cycle. Ratings can be achieved at the planning phase for communities, during the design, construction or fit-out phase of buildings, or during the ongoing operational phase. Green Star assesses a project based on a number of credits in various categories. A rating is awarded based on the percentage of available points that a project gains, and an overall score is assigned based on which platinum, gold, silver or a simple certification is granted.

In terms of certification processes, Green Star and BREEAM have similar approach. The trained assessor assists the design team in developing and documenting the sustainable design initiatives to achieve the desired rating and submits the documentation to the authorities whose panel validates the assessment and issue the certificate. The Accredited Professionals are appointed to the design team early in the design process. The assessor assists the design team in developing and documenting the sustainable design initiatives to achieve the desired rating and submits the documentation. In order to maintain the Green Star and BREEAM Accredited Professional qualification, the assessors must earn points through continuous professional development every year. While LEED does not require training, there is a credit available if an Accredited Professional is used [3]. The role of the Accredited Professional is to help gather the evidence and advise the client. The evidence is then submitted to the US Green Building Council (USGBC) [37] which does the assessment and issues the certificate. While LEED is dominated by the American ASHRAE standards, BREEAM takes its cue from European and UK legislation [3]. A computational simulation study carried out to quantitatively benchmark the three schemes showed that the case study office building received a high energy rating score in the Green Star scheme, but a low energy rating in the BREEAM scheme and it failed to be certified in the LEED scheme [32]. Also, the HVAC system was found to be the most heavily weighted variable in the energy assessment of the three schemes [32].

4 Adoption Rate

The adoption of rating schemes depends on various factors such as the energy policies and supporting mechanisms in respective countries. Other than the high-end office blocks of Australia's central business districts, the pace of progress in the adoption of Green Star rating has been low [39]. Mid-tier office buildings found all

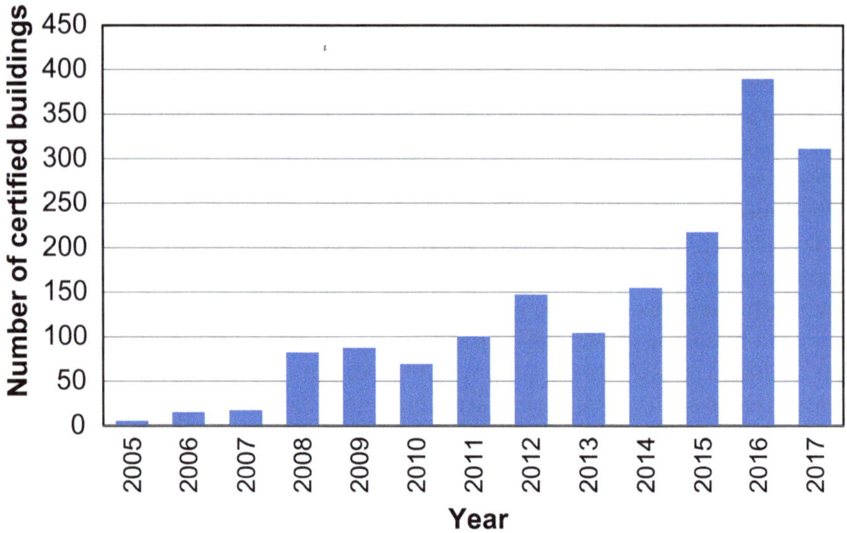

Fig. 1 Number of Green Star-certified buildings

across Australia including fringe areas, suburban centres and regional towns make up a significant proportion of Australia's overall commercial office building stock, but generally not Green Star rated. Figure 1 shows the adoption rate of Green Star buildings since 2005 using the data from ABS [1] and GBCA [15]. The GBCA website maps out the geographic location of the certified buildings. In 2005, there were four Green Star-rated buildings developed. This increased significantly to 83 buildings in 2008 moving to 400 buildings by 2016. Even though the numbers continue to grow and there are now 1712 certified buildings [15], they represent only a small percentage of total building stock. New South Wales has the largest number of certified buildings followed by Victoria and then Queensland and Western Australia.

Both the UK and USA seem to have better success with the construction industry adopting environmental rating schemes faster. As at the end of 2014, overall green building adoption rate in the USA was 13.11% of total numbers of commercial buildings [20]. The top six markets in the US green building sector has total percentage of office buildings built using either LEED or Energy Star above 19%, with the highest being Atlanta with 29.03%. The results show Energy Star being the preferred method to use when certifying an office building. The Energy Star rating system has had better success, growing from 2 to 9.69% during the same period [20] and 10.3% in 2016. The Energy Star programme expanded slightly in 2016, with 10.3% of all commercial office buildings in the largest markets now certified, up from 9.9%. LEED rating system has grown from 0.14% in 2005 to 5.32% in 2014. At the end of 2016, LEED certifications represented 4.7% of the total number of commercial office buildings across the 30 largest US office markets, up from 4.6% the year before.

In 2016, more than 7500 commercial buildings earned the Energy Star, bringing the total certified numbers to 29,500. The percentage of commercial office space that has been certified as "green" or "efficient" now stands at 38% across 30 office markets in the USA. That percentage grew from less than 5% in 2005. Large geographic variation in the adoption of LEED and Energy Star certification remains. For both LEED and Energy Star certification, the top three markets in terms of green building adoption by percentage of square footage are Chicago, San Francisco and Atlanta, with Chicago taking the top position for the first time. It is important to look at the progress these states have had with their rating systems and compare it to Australia to determine how to improve the adoption rate of green buildings.

The UK BREEAM programme's adoption rate has gone from 1.42% in 2003 to 5.9% in 2008 [7]. By 2012, 6739 commercial buildings were certified under BREEAM [2]. There are more than 2300 certified BREEAM projects in the UK. Around 7% of the nearly 7000 BREEAM-certified non-domestic buildings are in the retail sector, ranging from single units to entire shopping centres.

5 Cost

The cost of implementing certification is considered an important factor in the adoption rate. Like any other service, price is an unavoidable issue when putting the certification scheme into real practice on a large scale. When people pay the expert for the certificate, they will question themselves about the added value of that service [34]. The price of the certification is mainly dependent on the cost of the expert's work; consultant fees usually prevail over the administration fees. Some countries have a part of the price fixed, which corresponds to the mandatory fees paid for the issuing of the certificate in the system. In new buildings, the price is highly influenced by the methodology used and the geometry of the building. For existing buildings, the experience of the expert is the most noticeable factor. BREEAM fees are determined irrespective of the project size, whereas LEED and Green Star fees increase with project size. For Green Star, the certification fee schedule varies depending on the type of project, project's contract value and GBCA membership status. Historically, consulting costs to prepare the submission are in the order of A$100,000, including the Environmentally Sustainable Design (ESD) consultant and additional work required for the architects and consultants [18].

As noted by Santos and Whittchen (2011) in Europe, differences between the prices of Energy Performance Certification in the member states are more evident in the case of non-residential buildings, ranging from a couple of thousand dollars for small and simple buildings, up to A$30,000 per certificate for large and/or complex buildings. The price is often higher for the existing than for the new non-residential buildings, and the difference is more noticeable in the case of large and/or complex building.

Table 2 compares the cost of certification for the three rating systems. In order to compare the costs, a five star-certified building from GBCA website was selected for

Table 2 Comparison of certification cost

	Green star	LEED	BREEAM
Consultation cost (A$)	Up to 100,000 for ESD consultant and additional work required for the architects and consultants	NA	Approximately 31,000 for large buildings
Registration fees (A$)	Not applicable	1250 for members and 1700 for non-members	4000–6000
Certification fees (A$)	12,500–40,000	Up to 30,000	4000–6000
Estimated certification cost for a building with GFA 11,500 m^2 (A$)	18,500 for GBCA members and 23,500 for non-members	9000	6000

analysis. The selected building with a gross building area of 11,500 square metres, valued A$20 million. Standard certification fee for this building is A$18,500 for GBCA members and A$23,500 for non-members. Certification costs for a similar building using the LEED system are US$6470 (A$8998) and BREEAM system is around £3700 (A$6000) which are significantly lower compared to Green Star.

BREEAM is such a part of UK building certification that it is largely embedded into the regulations through Building Research Establishment (BRE), but unlike LEED, it is no longer administered by a non-profit organization. This has led some in the industry to criticize the programme, as it tends to charge significantly higher fees than LEED for one-off assessments [11]. BREEAM has licensed assessors who examine the evidence against the credit criteria and report it to BREEAM's parent company, BRE. BRE then assess the report and issue the certificate if it meets their requirements. Assessment is a two-stage process, as design stage (using documentary evidence) and post-construction (using site records and visual inspection). LEED, on the other hand, does not collect the evidence, the design team does. They then send the data to the USGBC, who examine it and issue the certificate if it meets their demands [11]. Across Europe, the price of certificates varies, due to the different economic realities in each country, and different methodologies used by those countries.

Even though the cost of certification is significant as its own, it is only up to 0.6% of the total building cost. The cost of construction can vary significantly. Generally, a 4 or 5 star Green Star rating can be achieved with no additional cost, provided there is good ESD integration from the start [9]. Additional large pieces of infrastructure, such as photovoltaics, cogeneration, blackwater treatment, added to achieve credits can significantly increase project costs and are often required for a 6 Star rating [18]. In a survey of building professionals [23], 49% of participants surveyed believe there is a problem with the affordability of green buildings, and the ratings are aimed at high-end projects. In the USA, after surveying LEED policy administrators, Retzlaff [31] found the majority of respondents believed public awareness and education played a vital role in communities adopting LEED-certified buildings. It is widely acknowledged that cost is a key variable that drives the market for high-performance buildings; therefore, it is important for the government to adopt cost competitive market transformation strategies such as competitive price for high-end materials, equipment and systems.

6 Incentives

Incentives serve as an instrument that can ultimately drive sustainable development in the building sector. Financial incentives include direct grants, tax incentives, rebates and discounted development application fees which are the most common green building incentives provided by the government [27]. Non-financial incentives include floor-to-area density considerations, technical assistance, expedited permitting, business planning assistance, marketing assistance, regulatory relief, guarantee programmes and dedicated green management teams in building and planning departments [8]. Non-financial incentives such as expedited permitting or technical assistance save owners' time by mitigating risk and process issues. Governments mostly favour the provision of non-financial incentives because no direct costs are involved [29].

Sauer and Siddiqi [35] compared the impact of three different incentives (financial and administrative incentives, and density bonus) provided at the county level on the production rates of LEED-certified multi-unit residential buildings in the USA and found that density bonus (i.e. zoning ordinances), which allows projects to achieve a higher unit density, leads to the production of more LEED-certified multi-unit residential buildings. Administrative incentives such as expedited permitting have a more significant impact on the adoption of green building by owners than financial incentives, such as tax credits [8].

In 2009, the Australian Government allocated A$90 million towards the Green Building Fund, which was a one-off funding scheme. This fund was set up to assist commercial building owners to improve the energy efficiency of their buildings. In 2010, Low Carbon Australia Limited (LCAL) was set up for the public charitable purpose of preserving and enhancing Australia's natural environment by helping Australian business, government and households take action to increase energy efficiency and reduce carbon emissions [22]. The energy efficiency investment portfolio has since moved to The Clean Energy Finance Corporation that promotes energy efficiency and cost-effective carbon reductions. LCAL acts as a financial provider to help develop Environmental Upgrade Agreements (EUAs). An EUA is a tripartite agreement between a building owner, local council and a finance provider where the finance is levied at a special charge by the local council. Under an EUA, lenders provide finance to a building owner for environmental upgrades, with the local council then collecting the repayments through its rates system and passing them on to the lender [10]. These agreements are designed to promote environmental friendly retrofits and upgrades of existing buildings. They were introduced in New South

Wales in 2011, South Australia in 2012 and recently introduced in Victoria in 2015 as part of the Sustainable Melbourne Fund [16].

The type of incentives offered in the top six green building markets in the USA includes tax credits or incentives, greater floor–area ratio density, and expedited and reduced permit fees [25]. These incentives are also seen in other states including Virginia, Maine, New Mexico, Arizona and Washington. In the UK, taxes on non-domestic energy use were introduced by the Climate Change Levy (CCL) in 2001 and the Carbon Reduction Commitment Energy Efficiency Scheme (CRCEES) in 2010. The imposition of the CCL was accompanied by incentives for companies to invest in energy efficiency such as Enhanced Capital Allowances (ECAs) allowing businesses to invest in designated energy-saving plant and machinery, and voluntary Climate Change Agreements (CCAs) allowing eligible energy-intensive industries to receive up to 90% reduction in the CCL if they signed up to stretching energy efficiency targets agreed with government. In addition, feed-in tariffs which took effect in 2010 applies to small-scale generation of electricity using eligible renewable technologies [38].

7 Discussion

Green Star compares well in terms of scoring criteria and methodology as well as performance standards in comparison to LEED and BREEAM which are considered international best practice tools. However, the adoption rate of Green Star has not grown significantly compared to the other two rating tools. It is to be noted that both countries have larger population compared to Australia and their rating tools have been developed for longer time period compared to Australia. One of the reasons for lower uptake could be Australia's relatively low energy prices which diminishes the financial incentive to act especially for private buildings. Government policies have played a substantial role in promoting energy efficiency improvements. Supporting measures are needed to ensure that rating schemes impact on the targeted market. The impact can be increased by incorporating other complementary measures, including energy requirements in building codes and financial incentives. Both Australia and Europe have similar policies regarding the disclosure of energy ratings in large commercial buildings. The European Energy Performance Certificate, however, extends to all buildings that are for sale or lease. The recent inclusion of building with floor–area from 1000 to 2000 m^2 hopes to push the small-size building market.

In the USA, local and state governments utilize various incentives to encourage the use of LEED. These include tax incentives, expedited and reduced fees, and relaxation on building area density or building heights. Many financial and non-financial incentives including tax incentives, expedited and reduced permits, and lenience for building density or building heights are available in the USA. These incentives have been seen to be successful in promoting green buildings in the private sector [25]. The adoption of energy benchmarking laws is rapidly advancing across US cities, counties and states. In total, 23 cities, Montgomery County, and the state of California have now enacted laws requiring large privately owned commercial buildings to annually measure and benchmark their energy consumption, as well as to publish the resulting scores [6]. Evidence from the 30 largest US commercial real estate markets suggests that these benchmarking and transparency laws may contribute to increased adoption of environmental building certification [6]. However, these programmes could be in jeopardy under the "America First" budget plan by the Trump administration. The lack of new policies and incentives throughout Australia points to a sceptical and non-committed attitude adopted by the government which is clearly seen by the elimination of the carbon tax. Some states in Australia are beginning to adopt financial incentives to promote the retrofit of commercial buildings through the Environmental Upgrades Agreement. This was adopted in New South Wales in 2012 with South Australia following in 2013 and Victoria in 2015. As these incentives are in their infancy, it is yet to be seen if they have an impact on the adoption rate.

There are several barriers that may affect the construction industry's uptake of energy efficiency measures. They include cost, information gaps (as relevant information is not always available at the right time to the right people), split incentives, skills shortages and delay in project commencement due to regulatory activities. Mandatory implementation can increase the impact considerably, but may be difficult to implement for budgetary or political reasons [30]. While there have been many energy efficiency improvements, some markets have proved resistant to change. Market diffusion activities including information and training, financial incentives, and financing will help to develop strategies to address these barriers.

As per the Green Building Adoption Index published by CBRE [6], the uptake of green building practices in the 30 largest US cities continues to be significant, but the growth shows abatement. It may indicate that the most sophisticated owners of the high-end buildings have pursued and achieved certification. Oyedokun [28] notes this as an indication of a low or complete lack of financial motivation for further expansion of the green building sector and reports that rather than green premium, issues around corporate social responsibility and energy efficiency legislation have been the main drivers for the green building market. It is to be also noted that most of the statistics do not cover buildings that achieve a high performance but do not pay for the certification. In addition, standards that were considered innovative once are common practices nowadays.

8 Conclusion

Much progress has been achieved in energy-efficient and environmentally sustainable buildings over the last two decades, and various rating schemes have evolved in different countries. Even though Green Star rating system in Australia is similar in rating criteria and performance standards in comparison to other rating systems in the USA and the UK, the market penetration of this rating system falls behind others in terms of adoption rate. The success of any rating scheme will depend on how cost-effectively it can be achieved. In order to achieve further advancements, the rating scheme should progress with time and contribute quantifiably to the environmental targets of the country. Both external and internal incentives are important instruments for promoting green building. However, it is not clear which one is more effective. Commercial building owners may not be motivated to achieve rating because the costs are not transferable to buyers who are actually reaping the benefits. A collaborative effort by the government and private sector and agreement on appropriate incentives is significant towards promoting participation of the private sector. Stringent regulations and increasing the minimum requirements to drive the bottom of the market in conjunction with more education and awareness within various stakeholders of the construction industry are imperative for better adoption. Also, it is very important to appraise the effectiveness of current government incentives.

Acknowledgements The author would like to thank Gareth Gilhooley, a former student, Deakin University, for his contribution in collecting some information for this chapter. The author would also like to thank Gillian Armstrong from the University of Adelaide for providing some information about BREEAM certification cost.

References

1. ABS, Australian Bureau of Statistics Buildings Approvals Australia, June (2015). http://www.abs.gov.au/AUSSTATS/abs@.nsf/DetailsPage/8731.0Jun%202015?OpenDocument Accessed 01 July 2015
2. BRE Global The Digest of BREEAM Assessment Statistics. http://www.breeam.org/filelibrary/Briefing%20Papers/BREEAM-Annual-Digest—August-2014.pdf. Accessed 18 Aug 2015
3. BSRIA (2009) https://www.bsria.co.uk/news/article/breeam-or-leed-strengths-and-weaknesses-of-the-two-main-environmental-assessment-methods/
4. Bernardi E, Carlucci S, Cornaro C, Bohne RA (2017) An analysis of the most adopted rating systems for assessing the environmental impact of buildings. Sustainability 9(7):1226. https://doi.org/10.3390/su9071226
5. Berry S, Marker T (2015) Residential energy efficiency standards in Australia: where to next? Energ Effi 8(5):963–974
6. CBRE (2017) Green building adoption index. https://dimension.maps.arcgis.com/apps/MapJournal/index.html?appid=a835df78ffe64a92be638610132b9b95
7. Chegut A, Eichholtz P, Kok N (2014) Supply, demand and the value of green buildings. Urb Stud (Sage Publications, Ltd.) 51(1): 22–43
8. Choi C (2009) Removing market barriers to green development: principles and action projects to promote wide spread adoption of green development practices. J Sustain Real Estate 1(1):107–138
9. Davis L (2010) The road to green property, Version 2.0, June 2010
10. DELWP (2017) Department of environment, land, water and planning: environmental upgrade finance. https://www.energy.vic.gov.au/energy-efficiency/environmental-upgrade-agreements. Accessed 20 Dec 2017
11. Dolan-Del Vecchio E (2016) LEED verus BREEAM: understanding the differences, View the space. http://blog.vts.com/leed-vs.-breeam-understanding-the-differences. Accessed 13 Jan 2018
12. Ekins P, Lees E (2008) The impact of EU policies on energy use in and the evolution of the UK built environment. Energy Policy 36(12):4580–4583

13. European Commission (2017). https://ec.europa.eu/energy/en/topics/energy-efficiency/buildings
14. GBCA (2013) Green building council of Australia the value of green star—a decade of environmental benefits. https://www.gbca.org.au/resources/gbca-publications/the-value-of-green-star/. Accessed 9 Jan 2018
15. GBCA (2017) Green star project directory. http://www.gbca.org.au/project-directory.asp 12 Nov 2017
16. GBCA (2017) Green building council of australia improving existing buildings—what financial options are available? http://www.gbca.org.au/advocacy/resources/improving-existing-buildings-what-finance-options-are-available/. Accessed 05 Sept 2017
17. Green Building Council of Australia Green Star. https://www.gbca.org.au/green-star/. Accessed 10 Oct 2017
18. Hampton A, Clay L (2016) Non-residential building environmental rating tools—a review of the Australian market, environment design guide EDG 85 AH, Mar 2016
19. Janda (2009) World status of energy standard for buildings, ECEEE 2009 summer study • act! Innovate! Deliver! Reducing energy demand sustainably. http://www.eci.ox.ac.uk/publications/downloads/janda09worldwidestatus.pdf. Accessed 10 Sept 2017
20. Kok N, Holtermans R, Pogue D (2015) National Green Building Adoption Index. http://www.cbre.com/o/international/AssetLibrary/Green-Building-Adoption-Index.pdf. Accessed 8 Oct 2017
21. Lee WL (2013) A comprehensive review of metrics of building environmental assessment schemes. Energy Build 62(2013):403–413
22. Low Carbon Australia (2013) Annual Report of the Low Carbon Australia Limited, 2012–2013. https://www.cefc.com.au/media/106762/lowcarbon-annual-report-2012-13_lr.pdf. Accessed 2 Sept 2017
23. McGraw Hill Construction World Green Building Trends. http://www.worldgbc.org/files/8613/6295/6420/World_Green_Building_Trends_SmartMarket_Report_2013.pdf. Accessed 01 Sept 2015
24. Ministry of Environment. http://www.environment.gov.au/minister/frydenberg/media-releases/mr20170629.html. Accessed 2 Sept 2017
25. Minzhen D (2012) To promote green buildings in China: Lessons from the USA and EU. J Res Ecol 3(2):183–191
26. Moloney S, Horne R (2015) Low carbon urban transitioning: from local experimentation to urban transformation? Sustainability 7:2437–2453. https://doi.org/10.3390/su7032437
27. Olubunmi OA, BoXia P, Skitmore M (2016) Green building incentives: a review. Renew Sustain Energy Rev 59(2016):1611–1621
28. Oyedokun TB (in press) Green premium as a driver of green-labelled commercial buildings in the developing countries: lessons from the UK and US. Int J Sustain Built Environ (2017). https://doi.org/10.1016/j.ijsbe.2017.12.007
29. Perkins M, McDonagh J (2012) New Zealand local government initiatives and incentives for sustainable design in commercial buildings. https://researcharchive.lincoln.ac.nz/bitstream/handle/10182/6916/eres2012_57%20%281%29.pdf?sequence=1&isAllowed=y. Accessed 3 Sept 2017
30. Rajagopalan P, Leung CYT (2012) Progress on building energy labelling techniques. Advan Build Energy Res 6(1):61–80
31. Retzlaff RC (2009) The use of LEED in planning and development regulation: an exploratory analysis. J Plan Educ Res 29(1):67–77
32. Roderick Y, McEwan D, Wheatley C, Alonso C (2009) A comparative study of building energy performance assessment between LEED, BREEAM and Green Star schemes, Building Simulation conference. Glasgow. www.iesve.com/content/downloadasset_1220
33. Sanderford AR, McCoy AP, Keefe MJ (2017) Adoption of energy star certifications: theory and evidence compared, Building Research & Information. https://doi.org/10.1080/09613218.2016.1252618

34. Santos P, Wittchen KB (2011) The price of energy performance certificates, European Union. http://www.epbd-ca.org/Medias/Pdf/Price_of_energy_performance_certificates.pdf. Accessed 11 Aug 2017
35. Sauer M, Siddiqi K (2009) Incentives for green residential construction. In: Proceedings of the construction research congress 2009
36. Talberg A, Hui S, Loynes K (2016) Science, Technology, environment and resources section, Australian climate change policy to 2015: a chronology, Parliamentary library
37. USGBC (2018) United States Green Building council. https://new.usgbc.org/
38. Uk Government (2017) Feed-in tariffs: get money for generating your own electricity. https://www.gov.uk/feed-in-tariffs/overview. Accessed 2 Aug 2017
39. Van der Heijden J (2014) Green building revolution? Only in high-end new CBD offices. http://theconversation.com/green-building-revolution-only-in-high-end-new-cbd-offices-24535. Accessed 24 May 2017

Assoc. Prof. Priyadarsini Rajagopalan, B.Tech. Arch. Eng., M.Bld.Sc., Ph.D. is an associate professor with expertise in building energy efficiency and energy benchmarking, indoor environmental performance and urban microclimate.

An End-User-Focused Building Energy Audit: A High-Density Multi-residential Development in Melbourne, Australia

Jin Woo and Trivess Moore

Abstract This chapter aims to demonstrate a building energy audit process using a case study of high-density multi-residential modular development in inner Melbourne, Australia. An energy audit is essential to understand where and how energy is used in buildings and consequently to identify those areas where improvements can be made. It includes a series of activities such as pre-survey data collection, walk-through inspection, data collection, analysis of the data collected and formulation of energy efficiency solutions. Extensive data were collected including indoor condition monitoring, occupant feedback and utility usage. The occupant survey identified thermal discomfort in summer, reporting overheating, dry and stuffy conditions. Energy consumption in the case study building was found to be significantly less than the average consumption in the same suburb. Surprisingly, energy consumption was found to be more likely to be affected by housing tenure types than physical building conditions such as orientation and height. The impact of building materials on occupants and the provision of air conditioning systems in the individual unit need to be further researched to resolve overheating problems. It is recommended that not only the design and physical conditions of buildings but also the socio-economic status of building residents could be main factors to achieve a high level of energy efficiency in multi-residential buildings.

1 Introduction

The main approaches to exploring improving energy efficiency in buildings are via simulation or experiment-based research; approaches which aim to reduce the energy performance gap between design and performance. There seems to be a tendency to overlook the real building performance through recording and analysing actual utility data and occupant feedback due to the difficulties in long-term monitoring and

J. Woo (✉) · T. Moore
Sustainable Building Innovation Laboratory, School of Property Construction and Project Management, RMIT University, Melbourne, VIC, Australia
e-mail: jin.woo@rmit.edu.au

P. Rajagopalan et al. (eds.), *Energy Performance in the Australian Built Environment*, Green Energy and Technology, https://doi.org/10.1007/978-981-10-7880-4_6

measurement of relevant data. The purpose of an energy audit is to understand where and how energy is *actually* used in a certain building and to identify and prioritise the areas where improvements can be made, either for that building or future designs [9]. There is a need to address this research gap, particularly in the residential sector where energy analysis is underutilised, if we are to understand how to further improve the energy and sustainability performance of new and retrofitted buildings [2, 14, 17] (see also Chapter "Low-Energy Housing as a Means of Improved Social Housing: Benefits, Challenges and Opportunities" for a detached housing case study).

Energy analysis has more typically been applied in commercial building research [23], and numerous energy audit procedures have been developed by energy and building services' professional bodies including the International Energy Agency [8], the Chartered Institution of Building Services Engineers [5] and the American Society of Heating, Refrigerating and Air-Conditioning Engineers [1]. A common energy audit procedure for commercial buildings can be classified into three levels: walk-through analysis, energy survey analysis and detailed analysis of capital-intensive modifications. A preliminary energy use analysis is a prerequisite for any audit. Audits can range from whole building energy to targeted audits with a limited scope such as lighting-only audits, cooling tower or boiler assessment and tenant improvement projects [1]. However, there are fundamental differences between domestic and non-domestic buildings as housing occupants are the building managers themselves (or their landlord if renters) who could improve or deteriorate building performance, manipulating indoor environment [3]. Consequently, the occupant behaviour and their interaction with building systems make the prediction of building performance more complicated in domestic buildings [11, 20, 22].

Post-occupancy evaluation (POE) studies demonstrate that building occupants are one of the most sophisticated and sensitive instruments available for housing performance evaluation and they also need to know how to use all the aspects of their home [7, 18, 21]. A pre- and post-domestic refurbishment study using occupant feedback techniques by Gupta and Chandiwala [6] identified significant gaps between modelled and actual energy consumption such as poor quality of indoor air and daylight, low operating internal temperatures and problematic noise transmission. The authors concluded that with an improved understanding of why this performance gap persists, more suitable user-centred low-carbon refurbishment interventions can be developed including improving data collection and analysis from occupant feedback on building performance which will lead to a better understanding of the context of *why* certain energy consumption and occupant practices occur [6]. Despite the benefits of POE in domestic buildings, Powell, Monahan, and Foulds [16] state that POE is an initiative to reduce the energy performance gap; however, it seems less common for residential buildings due to the difficulty in accessing occupants and their home for physical monitoring and/or occupant survey.

The aim of this chapter is to present a comprehensive POE approach to conducting a building energy audit using a multi-residential building in Melbourne, Australia, as a case study. Energy consumption in a multi-residential building can be different from a single detached house due to uneven conditions of individual units. The orientation and height (level) of individual units can vary, and the individual units share internal building components such as walls, floors and ceilings with adjoining units. This end-user-focused building energy audit will look into the end-use energy consumption of individual units in the case study building.

2 Case Study: An Affordable and Sustainable High-Density Housing Complex

The case study building is a graduated three-to-nine storey apartment and retail complex constructed in 2011. It was designed to be (at least in the context of Australian buildings) a highly innovative mixed use and mixed tenure apartment, offering sustainable and affordable living (Fig. 1). The building has several sustainable design features including shared open space (internal courtyard), skylight windows and voids to maximise daylight and winter sun, openings towards the building core to increase cross-ventilation, a central gas boosted solar hot water system and water recycling treatment plant. The social dimension of sustainability has been emphasised with the introduction of landscaped shared open space and a hierarchy of privacy and access in building design (Fig. 2).

Project background

Location: East Coburg, VIC

Building type: High-density housing (197 apartments)

Year of construction: Nov 2011

Climate: mild (climate zone 6)

Passive design strategies

- Landscaped shared open space
- Skylight windows
- Voids
- Openings towards the core
- Gas boosted solar hot water system
- Water recycling treatment plant

Fig. 1 Project building. *Source* DesignInc *Photograph credit* ©Dianna Snape

Table 1 Types of units

	N	No. of Bed/Bath	Floor area (m^2)
Type 1	111	Studio, 1 bed + 1 bath	28–50
Type 2	18	1 bed + 1 bath + 1 study	54
Type 3	46	2 bed + 1 bath	54–65
Type 4	22	2 bed + 2 bath	53–60

In total, there are 197 units. All individual units are categorised into four types according to the number of bed/bathrooms. A majority of the units (N = 111, Type 1) are either 'Studio' or '1 bed + 1 bath', having a range of the internal areas from 28 to 50 m^2. Table 1 summarised the types of individual units.

The case study building is located 7 km from Melbourne's CBD in a local climate classified as a mild temperate (climate zone 6) in the National Construction Code (NCC) climate zones of Australia. Historical climate data shows that the mean maximum temperature reached 26.9 °C in February and the mean minimum temperature reached 5.6 °C in July in the local area (a 38-year statistics of 1979–2016) [4]. In Australia, the energy efficiency of buildings is regulated under the NCC so as to use energy more efficiently and reduce greenhouse gas emissions as discussed in Chapter "Urban Climates in the Transformation of Australian Cities". The compliance requirements can be achieved by either using software rating tools (e.g. AccuRate Sustainability, FirstRate5 and BERS Pro Plus) based on Nationwide House Energy Rating Scheme (NatHERS) or alternatively complying with all the relevant NCC deemed-to-satisfy provisions, where minimum allowable elemental R-values are prescribed.

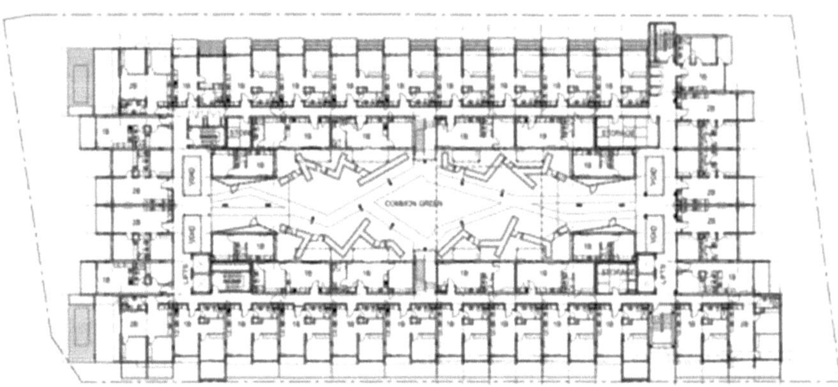

Fig. 2 Floor plan (Level 2). *Source* DesignInc

3 Approach

3.1 Pre-survey Data Collection

An environmental analysis report and a set of as-built plans were collected from the architect prior to conducting a physical building inspection. As part of the environmental analysis, a building thermal performance assessment was conducted using the FirstRate5 software. A sample of 18 apartment units was assessed, and an average of 6.17-star rating, ranging from 4.9 to 7.2, was achieved based on the NatHERS 0 (worst)-to-10 (best) star scale (see Chapter "Urban Climates in the Transformation of Australian Cities"). The case study building achieved a NatHERS rating beyond the minimum regulated building thermal performance when the building was designed.[1] The current building energy efficiency compliance is a 6-star rating in Australia, meaning that the annual energy consumption for space heating and cooling in homes is 114 MJ/m^2 in inner Melbourne.

3.2 Walk-Through Inspections

A series of inspections for the physical conditions and sustainable design features of the case study building were carried out in conjunction with utility reading on a monthly basis for a period of 12 months. The courtyard (Fig. 3) on the second floor not only gives natural ventilation and shading, but also provides the residents with an internal area for social catch-ups. During the initial inspections, minor damage was found (e.g. water damage to carpets in walkways), but it was repaired soon after the inspections.

Daylight which penetrates through skylight windows and voids into the building gives a reasonable amount of light into the building cores and corridors with the support of artificial lighting. Openings towards the building core, including operable windows and doors, help the residents control their environment. Interestingly, there were different ways of using the openings observed during the walk-through inspections. Some residents used the door openings for ventilation and others taped and sealed them, and even further the gap under the door was filled by a bunch of paper to block draught.

It was advised that the water recycling treatment system was not being used due to challenges completing the ongoing financing and maintenance of the system. Overall, the building had been well maintained, presenting a good standard of building cleanliness including building fabric and lighting fixtures.

[1] At the time of the buildings design, the requirement was to achieve a minimum of 5-star NatHERS rating.

3.3 Data Collection

A number of methods were adopted to collect building performance data including physical building condition monitoring and evaluation of occupant feedback and utility usage (Table 2). The primary focus of selecting data collection methods was to evaluate the energy and thermal performance of the case study building. The air infiltration rate was measured using a calibrated fan door blower test to examine ventilation heat loss and indoor air quality. Summer living room temperature and relative humidity were measured using HOBO data loggers. To ensure occupant thermal comfort, the Building User Satisfaction (BUS) survey, 'Housing Evaluation', was conducted. The questions of the Housing Evaluation survey include background, the residence overall, indoor conditions and personal control, lifestyle and utilities cost. Utility data including natural gas, electricity and water consumption were manually collected from all 197 apartments from March 2014 to February 2015 on a monthly basis. Despite the limitations of manual data collection such as recording frequency,

Fig. 3 Internal courtyard. *Source* DesignInc *Photograph credit* ©Dianna Snape

missed or lost readings and transcribing errors, the manual meter readings allowed the researchers to gain access to the building being monitored for monthly walk-through inspections of the case study building. There were also challenges in trying to engage more households to participate in the full range of monitoring, as such some tests had lower numbers than desired, e.g. blower door.

4 Results

This section presents the analysis of the occupant survey and energy consumption in the case study building. Although a series of building performance data were collected during the study, as the emphasis of this chapter is on user-focused, the analysis focuses on occupant feedback and the end-use of energy in the case study building. Additional analysis and impact from the case study building are reported elsewhere [12, 13, 24].

4.1 Occupant Survey

A POE survey was conducted using the BUS survey. The questions of the survey are generally measured on a seven-point Semantic differential using two adjectives with a neutral point (e.g. '1 = too cold and 7 = too hot'). The indoor conditions section includes thermal comfort, noise, lighting and personal control over the indoor environment, and a comment section is also provided after each question for further feedback. The format of the survey is a three-page hard copy, and the survey was distributed to all 197 apartments in the case study building via the post boxes in the foyers. A total of 28 households responded, representing a 14% response rate.

Thermal comfort of the individual units was evaluated over eight variables using a seven-point Semantic differential scale with a 'neutral' point of 4 which can be acceptable and comfortable for the respondents (Fig. 4). The respondents expressed overall satisfaction with their overall conditions (5.77) and thermal comfort (6.04) in winter. They, however, expressed dissatisfaction with their overall conditions (2.79)

Table 2 Summary of data collected

Method	Data	Sample (units)
Blower door test	Air infiltration rate	3
Indoor condition monitoring	Indoor temperature and relative humidity	6
Occupant comfort	Thermal comfort	28
Utility reading	Monthly gas, electricity and water usage	197

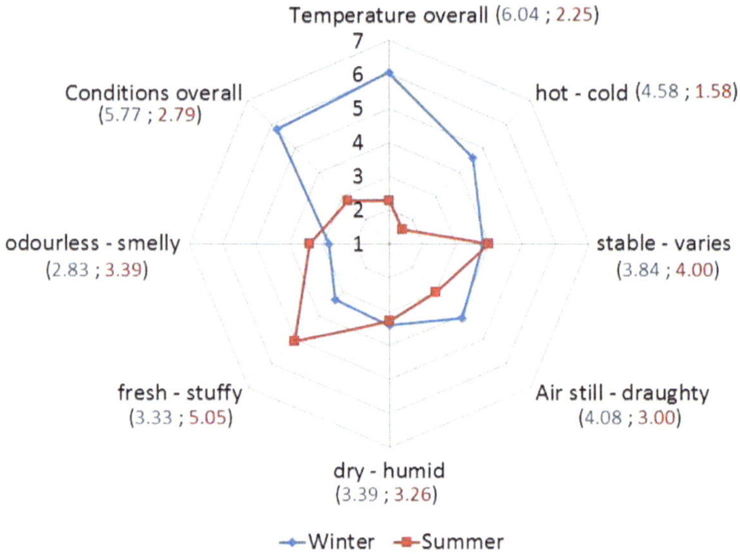

Fig. 4 Mean scores of thermal comfort—winter and summer

and thermal discomfort (2.25) in summer, reporting overheating (1.58), dry (3.26) and stuffy (5.05) conditions. This result could be interpreted: firstly, the overheated condition arose due to the higher local temperature in 2014 when the survey was conducted. The summer months from January to March in 2014 recorded an average 2.75 °C higher maximum temperature than the same months of the historical climate data. Secondly, this could be linked to passive design strategies such as cross-ventilation and operable openings integrated into the case study building with lack of active cooling systems of the individual units (unless the occupants installed these themselves). Lastly, it could be related to building materials and construction type as the building is a modular construction with engineered lightweight concrete floor and prefabricated building façade.

Personal control can be of potential importance to achieve energy savings and occupant thermal comfort. Leaman and Bordass [10] stated that it is vital to give occupants power of intervention to control their environment. Personal control over the indoor environment including heating, cooling, ventilation, lighting and noise was evaluated by the occupants over five variables using a 7-point Likert-type scale (Fig. 5). The respondent perceived a high level of control over heating and lighting, and a medium level of control over ventilation and cooling, and a low level of noise control. The results seem to be consistent with the building design and facilities provision of the case study building. A gas heater installed in the individual units enables them to maintain thermal comfort in winter, and lighting switches give them flexibility depending on their occupancy and behaviour. Operable windows, ventilation hatch and an electric ceiling fan also provide personal control over ventilation and

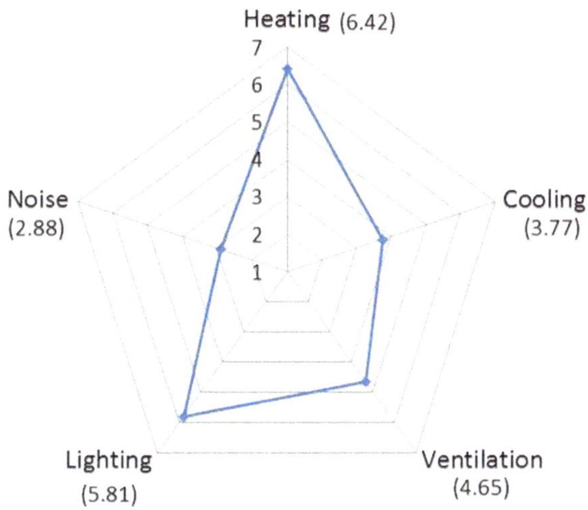

Fig. 5 Mean scores of personal control

cooling. Noise from outside such as traffic, nearby factory and communal garden, however, seems less controllable.

4.2 Energy Consumption

4.2.1 Monthly Consumption Patterns Per Household

A total of 358 days of the natural gas and electricity data were manually collected on a monthly basis. Due to the reading schedule differences, an average daily consumption per household was calculated based on the raw data collected and multiplied by the numbers of days for each month (Fig. 6). As natural gas is used as a main source for space heating, domestic hot water and cooking, the consumption significantly increased during winter months (July and August). The average annual consumption of natural gas was 1886.2 kWh per household (unit) and this is equivalent to 5.2 kWh/per household/day. The average annual consumption of electricity was 1350.1 kWh per household and this is equivalent to 3.7 kWh/per household/day. The average monthly pattern of electricity consumption per household does not show significant changes, indicating slight increase during winter (July and August) and summer months (Dec to Feb). It can be interpreted that active control systems such as air conditioners/electric fans and electric heaters/blankets were used during those months. A previous research conducted by NAGA [15] indicated an average daily gas usage of 140 MJ (or 38.9 kWh) and an average daily electricity usage of 11.7 kWh per household in the same postcode area. It is noted that the case study building in this chapter is a high-density apartment building with small internal

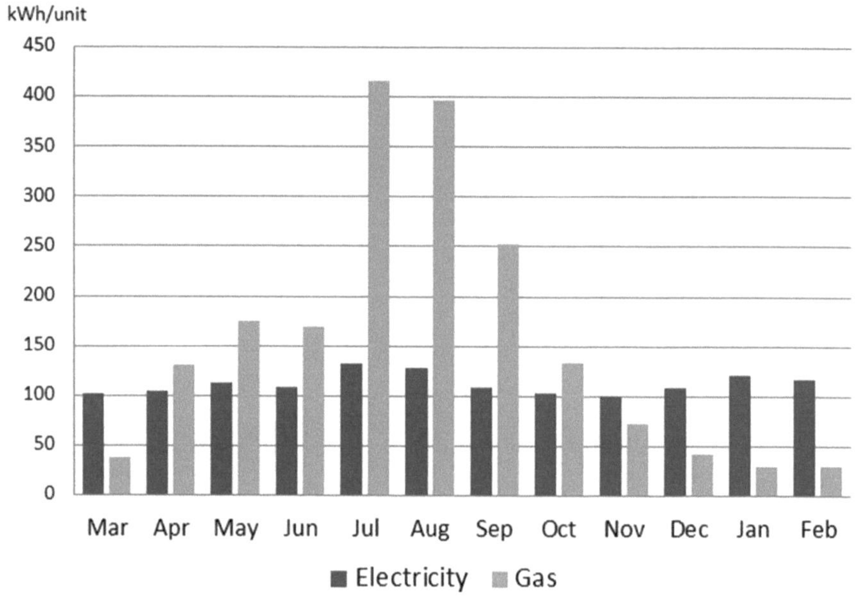

Fig. 6 Monthly patterns of energy consumption

space areas, whereas the majority of detached houses were analysed in the NAGA research. Thus, it seems more relevant to compare the daily consumption per person than per household. For example, the daily electricity consumption calculated from this study (3.7 kWh/hh/day) can be compared to the daily consumption per person (4.9 kWh/pp/day) in the same suburb from the previous research [15], as this case study building accommodates only one or two bedroom units with a single or couple occupants based on the occupant survey results.

4.2.2 Annual Energy Consumption Per Household

This section reports the analysis of annual energy consumption by physical conditions and tenure in the case study building. In order to detect any differences existed in energy consumption according to individual unit type, orientation, height (level) and tenure, One-way ANOVA test was conducted using SPSS package. One-way ANOVA assumes that data are normal and all groups share a common standard deviation even if they have different means. To ensure the homogeneity of variance in this analysis, Levene's test was used. If group sizes are vastly unequal and homogeneity of variance is violated, a nonparametric test was conducted. It is noteworthy that the raw data of 358 days with initial measurement units (m^3 for natural gas and kWh for electricity) were analysed in order to reduce errors.

Table 3 Analysis by unit type

	Type of unit	*N*	Mean	*F*-value	Sig.
Natural gas (m^3)	Type 1	111	126.6	4.113	.007
	Type 2	18	226.4		
	Type 3	46	195.8		
	Type 4	22	201.1		
	Total	197	160.2		
Electricity (kWh)	Type 1	107	1097.1	6.715	.000
	Type 2	18	993.4		
	Type 3	40	1422.0		
	Type 4	22	1500.6		
	Total	187	1204.1		

Natural gas consumption per household demonstrated a clear difference between Type 1 and Type 2. As expected, Type 1, having the smallest internal space area, consumed the least natural gas. Surprisingly, Type 2 showed the highest natural gas consumption among all types although Type 3 and 4 have more bedrooms and larger internal spaces than Type 2 (refer to Table 1 in Sect. 2).

It seems that the open plan of living–dining–study area of Type 2 requires more heating and cooling load from the heater and electric fan installed in the living space of the unit. On the contrary, Type 3 and 4 have more bedrooms with less open space. Electricity consumption showed a statistically significant difference between two groups: Type 1/2 and Type 3/4. This clearly represents the number of bedrooms, consequently the area of internal space, caused more electricity consumption. Table 3 summarised the natural gas and electricity consumption per household by unit type.

Surprisingly, both natural gas and electricity consumption per household showed no statistically significant difference according to orientation and height. Rather, the energy consumption per household showed a statistically significant difference in housing tenure types (Table 4). The case study building has three types of mixed tenure: privately owned, social housing through a single housing provider for low-income households (HGS) and affordable housing for low–moderate-income households (NRAS).

Over half of the units were privately owned in the case study building. Natural gas consumption per household demonstrated a statistically significant difference between tenure types based on a nonparametric test (Kruskal–Wallis test) due to the violation of homogeneity of variance. The residents of affordable housing (NRAS) were the largest consumer of natural gas followed by those of social housing (HGS). The residents of privately owned units, however, consumed natural gas the least. On the contrary, the private owners were the largest consumer of electricity per household in the same period followed by the residents of affordable housing (NRAS). It could be further examined based on their lifestyle such as occupancy and appliance usage.

Table 4 Analysis by tenure

	Tenure	N	Mean	F-value	Sig.
Natural gas (m^3)	Private	108	129.5	n.a.	sig.
	HGS	57	188.0		
	NRAS	31	216.3		
	Total	196	160.3		
Electricity (kWh)	Private	102	1290.3	3.491	.033
	HGS	57	1053.4		
	NRAS	27	1151.9		
	Total	186	1197.6		

[a]Natural gas consumption was analysed using a nonparametric test (Kruskal–Wallis test)

4.2.3 Energy Use Intensity (Per m^2)

Energy use intensity (EUI) is used to quantify and compare the operational energy consumed by buildings. It can be calculated by dividing the total energy consumed by the building in one year by the total gross floor area of the building. Again, to reduce errors, the raw data of 358 days with initial measurement units (m^3 for natural gas and kWh for electricity) were analysed in this section. An average natural gas use intensity was 3.05 m^3 per m^2 (or 118.0 MJ or 32.8 kWh), and an average electricity use intensity was 23.2 kWh per m^2 (83.5 MJ) during the 358-day study period.

No statistically significant difference was found in all physical characteristics including unit type, orientation and height; however, a statistically significant difference in both natural gas and electricity consumption was found between housing tenure types (Table 5). A nonparametric test (Kruskal–Wallis test) and One-way ANOVA support the same rank of natural gas and electricity consumption: the residents of affordable housing (NRAS) were the largest consumer of natural gas followed by those of social housing (HGS). The residents of privately owned units, however, consumed natural gas the least. Also, the private owners were the largest consumer of electricity per household in the same period followed by the residents of affordable housing (NRAS).

5 Discussions

One significant benefit of using a standardised survey seems to benchmark or compare a building's performance against similar buildings. The BUS methodology provides a building performance analysis against the international housing evaluation benchmark, and based on the international benchmark, the case study building overall performed well across assessment criteria except 'summer conditions' and 'noise'. This analysis seems consistent with the indoor condition monitoring that summer

Table 5 Utility consumption intensity (per m^2) by tenure

	Tenure	*N*	Mean	*F*-value	Sig.
Natural gas[a] (m^3 per m^2)	Private	108	2.4816	n.a.	sig.
	HGS	57	3.5899		
	NRAS	31	4.0441		
	Total	196	3.0511		
Electricity (kWh per m^2)	Private	102	24.9788	4.217	.016
	HGS	57	20.1178		
	NRAS	27	22.7732		
	Total	186	23.1695		

[a]Natural gas consumption was analysed using a nonparametric test (Kruskal–Wallis test)

overheating risks were identified during summer living room temperature monitoring. Thermal discomfort in summer was also identified through the occupant survey. Although various passive design strategies such as cross-ventilation and operable openings are integrated into this case study building, an electric ceiling fan was not sufficient to maintain the indoor temperature cool down in summer unless an air conditioner was installed in the individual unit. Furthermore, the case study building is a modular construction with engineered lightweight concrete floor and prefabricated building façade. The impact of building materials on occupant thermal comfort needs to be further researched to resolve overheating in this type of construction, although the building is seen as an exemplar of this more innovative (for Australia) construction method [13].

Energy consumption in the case study building tends to be far less than the average consumption in the previous research. However, a direct comparison between them does not seem to be relevant due to dwelling types [19].

The difference in energy consumption was found according to the individual unit type (Table 3). It clearly demonstrates that the internal space layout and heating and cooling systems of the individual unit influence energy consumption even in the same size of internal space. Unexpectedly, there was no difference found in energy consumption according to the orientation and the height (level) of the individual unit. Rather, housing tenure types are more likely to influence energy consumption in the case study building. Further analysis of energy consumption from a socio-economic perspective would be required to fully understand the residents and to provide better design strategies in order to achieve building energy efficiency in multi-residential buildings.

The residents were asked about the utility costs compared with their previous accommodation as part of the Building User Satisfaction (BUS) survey. Three questions about heating, electricity and water were assessed using a seven-point Likert-type scale (1 much lower and 7 much higher). The respondents expressed spending lower utility costs for heating (3.13), electricity (3.26) and water (2.83) compared to the utility costs spent in the previous accommodation.

Extensive data were collected using a multidisciplinary approach in the case study building to provide a more holistic understanding of the building performance and occupants' perspectives. Although the on-site energy audit was straightforward once the researchers gained access to the case study building, the building user survey was not as simple. The initial response rate to the posted survey was only five per cent and a second mail-out was distributed via the post boxes to improve the response rate. While more residents responded, overall numbers engaging in the research were lower than desired. Alternative ways of occupant engagement are required to ensure that high-quality data can be collected. For example, having a communal event to promote the research project and remind the residents to complete the survey might be an effective strategy if the owners' corporation can be supportive.

6 Conclusion

This chapter demonstrates an end-user-focused building energy audit process using a case study of high-density multi-residential development in inner Melbourne. This process includes a series of activities such as pre-survey data collection, walk-through inspection, data collection, analysis of the data collected and formulation of energy efficiency solutions. Extensive data were collected including physical building condition monitoring, occupant feedback and utility usage. The occupant survey identified thermal discomfort in summer reporting overheating, dry and stuffy conditions. The passive design strategies such as cross-ventilation and operable openings integrated into the case study building need to be closely examined in conjunction with local climate and occupant behaviour. Furthermore, the impact of building materials on occupant thermal comfort and the penetration of air conditioning systems need to be further researched to resolve overheating problem. Energy consumption in the case study building seems far less than the average consumption in the same postcode area, which could result from differences in the dwelling types. Surprisingly, energy consumption is more likely to be affected by housing tenure types than physical building conditions such as the orientation and height (level) of individual units. Further analysis of energy consumption from a socio-economic perspective would be a strategy to achieve building energy efficiency in multi-residential buildings.

References

1. ASHRAE (2011) Procedures for commercial building energy audits, 2nd edn. ASHRAE, Atlanta
2. Berry S, Whaley D, Davidson K, Saman W (2014) Do the numbers stack up? Lessons from a zero carbon housing estate. Renew Energy 67:80–89. https://doi.org/10.1016/j.renene.2013.11.031
3. Berry S, Whaley D, Davidson K, Saman W (2014) Near zero energy homes—what do users think? Energy Policy 73:127–137. https://doi.org/10.1016/j.enpol.2014.05.011

4. Bureau of Meteorology (2016) Climate data online. Retrieved from http://www.bom.gov.au/climate/data/index.shtml
5. CIBSE (2008) Energy Benchmarks—CIBSE TM46: 2008
6. Gupta R, Chandiwala S (2010) Understanding occupants: feedback techniques for large-scale low-carbon domestic refurbishments. Build Res Inf 38(5):530–548. https://doi.org/10.1080/09613218.2010.495216
7. Hadjri K, Crozier C (2009) Post-occupancy evaluation: purpose, benefits and barriers. Facilities 27(1/2):21–33. https://doi.org/10.1108/02632770910923063
8. IEA (2013) Total energy use in buildings. Analysis and evaluation methods. Final Report Annex 53. Retrieved from Japan http://www.iea-ebc.org/Data/publications/EBC_Annex_53_Main_Report.pdf
9. Ingle A, Moezzi M, Lutzenhiser L, Diamond R (2014) Better home energy audit modelling: incorporating inhabitant behaviours. Build Res Inf 42(4):409–421. https://doi.org/10.1080/09613218.2014.890776
10. Leaman A, Bordass B (2001) Assessing building performance in use 4: the Probe occupant surveys and their implications. Build Res Inf 29(2):129–143. https://doi.org/10.1080/09613210010008045
11. Menezes AC, Cripps A, Bouchlaghem D, Buswell R (2012) Predicted versus actual energy performance of non-domestic buildings: Using post-occupancy evaluation data to reduce the performance gap. Appl Energy 97:355–364. https://doi.org/10.1016/j.apenergy.2011.11.075
12. Moore T (2018) Strategic niche management and the challenge of successful outcomes. In: Moore T, de Haan F, Horne R, Gleeson B (eds) Urban sustainability transitions. Australian Cases—International Perspectives. Springer, Singapore
13. Moore T, Higgins D (2016) Influencing urban development through government demonstration projects. Cities 56:9–15. https://doi.org/10.1016/j.cities.2016.02.010
14. Moore T, Ridley I, Strengers Y, Maller C, Horne R (2017) Dwelling performance and adaptive summer comfort in low-income Australian households. Build Res Inf 1–14. https://doi.org/10.1080/09613218.2016.1139906
15. NAGA (2012) Moreland energy profile 2004–2012. Retrieved from http://www.naga.org.au/uploads/9/0/5/3/9053945/naga_-_mep_-_moreland_2012_final.pdf
16. Powell J, Monahan J, Foulds C (2016) Building futures: managing energy in the built environment. Routledge
17. Pretlove S, Kade S (2016) Post occupancy evaluation of social housing designed and built to Code for Sustainable Homes levels 3, 4 and 5. Energy Build 110(Supplement C):120–134. https://doi.org/10.1016/j.enbuild.2015.10.014
18. Stevenson F, Rijal HB (2010) Developing occupancy feedback from a prototype to improve housing production. Build Res Inf 38(5):549–563. https://doi.org/10.1080/09613218.2010.496182
19. Strata Community Australia (n.d.) Smart blocks. Retrieved from http://smartblocks.com.au/
20. Sunikka-Blank M, Galvin R (2012) Introducing the prebound effect: the gap between performance and actual energy consumption. Build Res Inf 40(3):260–273. https://doi.org/10.1080/09613218.2012.690952
21. Turpin-Brooks S, Viccars G (2006) The development of robust methods of post occupancy evaluation. Facilities 24(5/6):177–196. https://doi.org/10.1108/02632770610665775
22. Van den Brom P, Meijer A, Visscher H (2017) Performance gaps in energy consumption: household groups and building characteristics. Build Res Inf 1–17. https://doi.org/10.1080/09613218.2017.1312897
23. Woo J (2014) A systematic post-occupancy evaluation in green-rated high-rise office buildings. Paper presented at the Across: Architectural Research through to Practice: 48th International Conference of the Architectural Science Association 2014, Genoa, Italy
24. Woo J (2016) The energy use and occupant comfort of a high-density multi-residential development. Paper presented at the 32th International Conference on Passive and Low Energy Architecture (PLEA 2016), Los Angeles, United States

Dr. Jin Woo BS, MS, MPD, Ph.D., Green Star AP is a lecturer in building science with expertise in post-occupancy evaluation of buildings.

Dr. Trivess Moore BSc. (Env.), Ph.D. is a research fellow with expertise in energy efficiency, renewable energy, the built environment and transitioning to a low-carbon future.

Low-Energy Housing as a Means of Improved Social Housing: Benefits, Challenges and Opportunities

Trivess Moore

Abstract Rising energy costs are significantly impacting low-income households. These households can struggle to pay their utility bills, and/or self-ration how much energy they consume which impacts on liveability within the home, such as the provision of appropriate thermal comfort. While incremental progress is being made in terms of improving the energy efficiency of housing in many developed countries, such improvements are typically inaccessible to low-income or social housing tenants. This chapter presents outcomes of a multi-year evaluation of a cohort of low-energy social housing from Horsham in regional Victoria, Australia. The analysis includes technical performance data and is supplemented with the occupants' own stories about improved liveability outcomes. It is clear that the evidence supports aspirations by the state housing agency, which owns and maintains the housing, to move beyond their current minimum housing standards for new construction. A combination approach, whereby the thermal performance of the dwelling is improved, in addition to including renewable energy generation, will address several goals of social (or public) housing providers—namely improving quality of life, health outcomes, finances and poverty. In addition, such housing will help them achieve organisational or broader government sustainability goals such as reducing greenhouse gas emissions and fossil fuel energy consumption.

1 Introduction

The unsustainable energy performance of housing in Australia, and many developed countries, is not just an issue for the environment (see Chapters "The Built Environment in Australia", "An End-User Focused Building Energy Audit: A High-Density Multi-Residential Development in Melbourne, Australia", "Are we Living with Our Heads in the Clouds? Perceptions of Liveability in the Melbourne High-Rise

T. Moore (✉)
Sustainable Building Innovation Laboratory, School of Property Construction and Project Management, RMIT University, Melbourne, VIC, Australia
e-mail: trivess.moore@rmit.edu.au

P. Rajagopalan et al. (eds.), *Energy Performance in the Australian Built Environment*, Green Energy and Technology, https://doi.org/10.1007/978-981-10-7880-4_7

Apartment Market" and "The Way Forward-Moving Towards Net Zero Energy Standards" from this book). Sustainable housing is increasingly also about improving outcomes for occupants in the dwelling, by improving thermal comfort, liveability and reducing costs of living [8, 9, 28]. This last point is of increasing concern in many countries with the cost of energy rising significantly in recent years. In Australia, from 2007 to 2017 the price of electricity rose by 62% and gas 71% (both inflation adjusted) [5]. Other countries have experienced varied rises (and falls) in the price of energy, for example in the 10 years to 2016 the price of electricity rose by almost 21% in the USA [14] but rose by 65% in England [12].

It is low-income households who are most at risk from increasing energy prices. For this chapter, low-income households refer to those households who are in the lowest equivalised disposable household income quintile as defined and measured by the Australian Bureau of Statistics [1]. These households typically have limited ability to absorb additional costs which can result in disconnection of utilities when payments are not made on time [2, 6, 10, 11]. In 2015–16, there were more than 135,000 energy disconnections in Australia demonstrating the size of the problem [4, 15]. There is evidence that some low-income households reduce their energy costs by self-rationing their energy consumption which can lead to other issues such as compromising appropriate thermal comfort levels [10, 16, 18, 23]. Research has found that some low-income households will make trade-offs from other areas of their life (e.g. healthy eating, healthcare, education) to ensure they can pay their energy bills [10].

While some progress is being made in terms of improving the energy efficiency across the residential sector in Australia (e.g. the 6 star National House Energy Rating Scheme (NatHERS) requirement for new housing or retrofit of existing housing, see Chapter "The Built Environment and Energy Efficiency in Australia: Current State of Play and Where to Next"), such improvements are typically inaccessible to low-income social housing tenants who have limited control over what dwelling they live. Social housing can be a mix of age and quality as government and not-for-profit social housing providers juggle the need to provide more housing as well as upgrading their existing housing stock with challenges including capital costs, split incentives and conflicting or complex information [2, 6]. Occupants in social housing generally have limited means to make improvements themselves and often have older, less energy efficient appliances (e.g. fridge, washing machine) [2].

While there are increasing numbers of sustainable housing projects occurring around the world, there are less which are specifically targeted at social housing [17, 19, 25, 26, 29]. This chapter presents a case study of a multi-year evaluation of a cohort of low-energy social housing from Horsham, Victoria.

2 Social Housing in Australia

This section provides a brief overview of social housing in Australia. As defined by the lead social housing provider in Victoria, Australia, social housing is made up of two types of housing, public housing (state owned/managed) and community housing (not-for-profit owned/managed) [13]. It is for people on low incomes who need housing, including those who have recently experienced homelessness, family violence or have other special needs and can be for short- or long-term accommodation.

There are gross income thresholds set to qualify for social housing which differs between states and organisations around Australia. For example, in Victoria for 2017 the state government housing provider has a threshold range of $981 gross weekly income for a single occupant household to $2025 gross weekly income for a family with dependent children. There are also asset limits which also apply. Households which qualify for social housing are provided access to housing at below-market rental rate value and may be provided with additional financial or other assistance to help them meet minimum quality of life requirements. In Victoria, where this chapter is focused, the state government housing provider sets the rent cap for low-income households at 25% of their gross income.

Information by the Australian Institute of Health and Welfare [3] reported that in 2016 there were more than 845,000 tenants living in 394,000 social housing dwellings across Australia. Almost 80% of these were through state government housing provision (i.e. public housing). Tenants are more likely to be older persons over the age of 55 years or children under the age of 15 years. Almost two-thirds of tenants are women and just over half of all households are single adult households. Approximately 41% of households in public social housing have been in their tenancies for more than 10 years. Social housing covers a range of housing types from apartments through to detached housing.

The providers of social housing often face complex and sometimes competing objectives which must be balanced out. For example, one typical objective is to provide housing for all those in need. In Australia, there is a need for more social housing and so there is an ongoing requirement to add additional houses to keep up with demand. Developing new housing, or purchasing existing housing, is a costly exercise and so due consideration must be given to ensure that the best value for money is achieved.

3 Pushing Design and Sustainability Boundaries

The Department of Health and Human Services (the Department) is a Victorian State Government Department which provides, amongst other services, housing to low-income households in Victoria. The Department currently has a portfolio of more than 84,000 dwellings which they own and manage. Their portfolio of housing includes different dwelling types (e.g. detached housing, apartments), locations (urban and regional) and cater for a range of different living arrangements (e.g. single occupants, family living, special needs, elderly), highlighting the complexity they face in providing housing for those in need.

Within the broader context of the government's requirements for improved sustainability outcomes, the Department has been exploring how to improve the sustainability and performance of their dwellings (both the physical building and how they are being used by tenants) and what impacts on social and health outcomes are likely to be for their tenants, as well as contributing to the governments broader sustainability improvement goals. For example, the Department has been involved in developing higher density apartments such as K2 which included passive design features, rainwater harvesting, grey water reuse, solar hot water and photovoltaics for renewable energy generation [30]. The K2 apartments were designed to have improved performance compared to standard apartments at the time of its construction. This included using 55% less mains electricity, 46% less mains gas and 53% less mains water.

However, the Department recognised that this was just the beginning and more needed to be done in relation to improving the performance and sustainability of their building stock, including developing a plan for lower density housing [24]. The Department made a strategic decision to develop an innovative and leading sustainable social housing exemplar project which went significantly beyond minimum standards to explore what the costs and benefits were for both the Department and for the tenants, and how the development could inform future departmental housing developments and standards.

Horsham, in regional Victoria, was selected as the location for the development as it offered extreme summer and winter climatic conditions (climate zone 27 in NatHERS—hot, dry summer, cool winter). This allowed for comprehensive analysis of how such housing performs in the context of a changing physical climate, with the predictions that Victoria (like other locations) will be facing more frequent and severe weather events.

The result was the construction of four two-bedroom, single storey, sustainably designed units with a NatHERS rating of 8.9 stars (referred to herein as low-energy houses or LEH). These low-energy houses have a predicted heating and cooling energy load of 25 MJ/m^2/year and utilised a number of key design and technology features to achieve the low-energy outcome such as improved insulation, glazing and thermal mass (see Table 1; Figs. 1 and 2). Seven control dwellings and households were also included in the research. The control houses were all located in Horsham and built at a similar time to the low-energy houses; however, they were built to the Department standard requirements at that time; a 6 star NatHERS rating with a predicted heating and cooling energy load of 108 MJ/m^2/year, but going beyond this minimum requirement by also including solar hot water and a rainwater tank not plumbed into the house (see Fig. 3). The design elements for both the low-energy and control houses are listed in Table 2.

The additional capital cost for the sustainability elements of the low-energy houses was calculated to be $75,800 per dwelling (see Sect. 4.3). The Department also provided each low-energy household with a manual on how to maximise the performance of their new dwellings and conducted a 2-hour hands-on house tour to show tenants how the houses operated before they moved in to ensure that all residents understood the various sustainability design elements and technologies included in the dwelling.

Table 1 Design and technology inclusions for the low-energy and control houses

Low-energy house	Control house
8.9 Stars—25 MJ/m^2/year predicted heating and cooling energy load	6.0 Stars—108 MJ/m^2/year predicted heating and cooling energy load
Solar hot water systems (gas boosted)	Solar hot water systems (gas boosted)
Two 5000L rainwater tanks shared between the houses and plumbed into toilets	Basic rainwater tanks (not plumbed into the house)
Passive solar design	
Optimum orientation	
Advanced roof design	
Improved levels of ceiling/wall/floor insulation	
External window shading	
Access to natural ventilation	
Increased thermal mass	
Reverse brick veneer construction on back half of housing	
Improved glazing	
1.5 kW solar photovoltaics (PV) system per house, with a 60c/kWh feed-in tariff	

Fig. 1 Picture of one of the low-energy houses in 2012. *Source* Trivess Moore

RMIT University was engaged to conduct a post-occupancy evaluation which began at the end of the first year of the low-energy houses being occupied in April 2013 and went until October 2015. The methods included:

- Three separate rounds of in-home interviews with householders across three years;

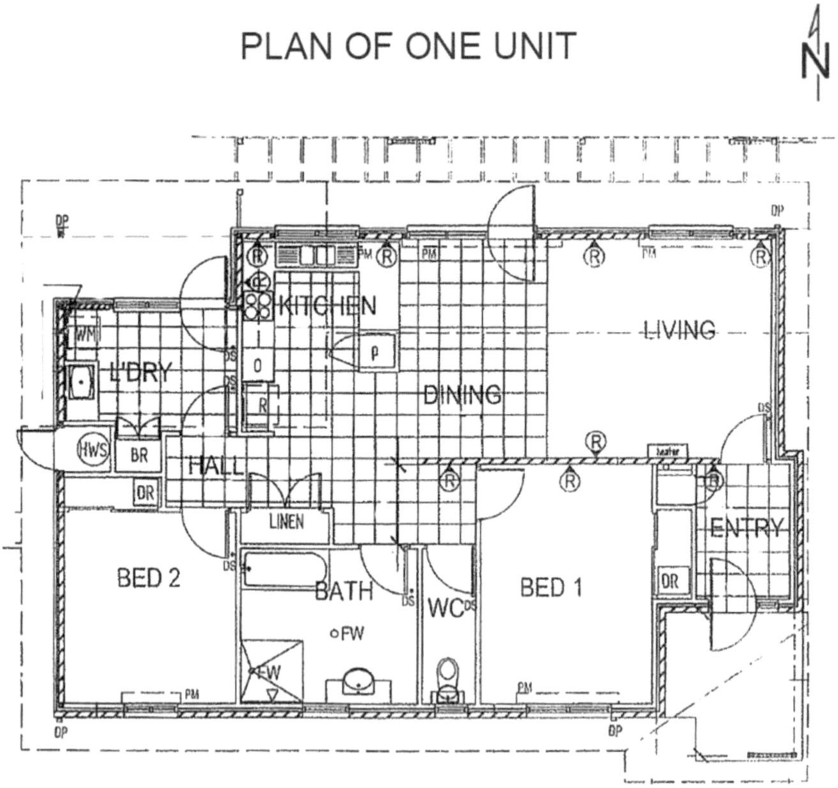

Fig. 2 Floor plan for one of the low-energy houses (while the low-energy house plans were almost identical, there were significant differences between the layout of the control houses so no example floor plan has been included.) *Source* Department of Health and Human Services

- Two rounds of interviews with key stakeholders involved in the conception, design, construction and ongoing management of the low-energy houses. This included the architect, building, electrician as well as key regional and head office Department stakeholders;
- A housing performance and cost–benefit analysis.

For both sets of houses, utility consumption (electricity, gas and water) and renewable energy generation data (where relevant) were monitored at 15-minute intervals via in-home monitoring equipment. This monitored data was cross-checked with utility billing data to improve accuracy. Hobo loggers were also used to measure temperature and humidity data throughout the study for the main living area and main bedroom in the dwellings.

A third-party engineer engaged by the Department to set up the data monitoring and initial utility consumption models, Organica Engineering, developed a Department "standard" performance scenario assuming a two person "average" occupancy

Fig. 3 Picture of one of the control houses in 2012. *Source* Trivess Moore

living in housing meeting the current Department standards at that time. This scenario was applied for comparison to the 11 case study dwellings (four low-energy houses and seven control houses).

Further, details of the methods and outcomes can be found in the detailed project report [24].

4 Analysis

This section provides analysis on how the low-energy houses performed across their first three years of occupation. While there are broader elements of sustainability included in these dwellings (e.g. water efficiency measures), the focus of this chapter is on the energy performance (and by association the thermal performance).

4.1 Energy and Environmental Performance

The low-energy houses improved energy performance through the inclusion of renewable energy technology as well as improving the thermal performance of the

Table 2 Characteristics of each household and house

Code	Household makeup and approximate age at first interview	Dwelling star rating	Thermal performance (heating and cooling MJ/m^2 annum)	Total internal area (m^2)	Total internal area conditioned (m^2)	Cooling technologies used in house
Low-energy houses						
LEH-A	Couple (early 20s) with two children (aged 3 and under)	8.9	26	100	74	Two ceiling fans in living area
LEH-B	Older couple husband and wife (60+ years)	8.9	25	99	72	Two ceiling fans and split-system reverse cycle air conditioner in living area
LEH-C	Single woman (60+ years)	8.9	26	100	74	Two ceiling fans in living area
LEH-D	Single woman (55+ years)	8.7	33	99	73	Two ceiling fans in living area
Control houses						
ConA	Husband and wife (60+ years)	6.0	108	82	73	Split-system reverse cycle air conditioner
ConB	Single mother (mid 20s) and child (3 years old)	6.0	108	97	84	None
ConC	Single male (45 years)	6.4	98	52	40	Pedestal fans
ConD	Husband and wife (55+ years) and teenage boy	6.0	108	97	84	Portable air conditioner
ConE	Single mother (50+ years) and teenage daughter	6.0	110	88	75	Wall unit air conditioner
ConF	Husband and wife (65+ years)	6.0	108	85	76	Split-system reverse cycle air conditioner
ConG	Householder information unknown, monitored performance data only					

Table 3 Summary of average annual utilities consumed/generated from each dwelling from June 2012 to May 2015

	Electricity consumed (kWh)	Total electricity consumed per m^2 of dwelling (kWh)	Electricity bought (kWh)	Total electricity bought per m^2 of dwelling (kWh)	Renewable energy generated (kWh)	Gas consumed (MJ)	Total gas consumed per m^2 of dwelling (kWh)	Total energy consumed (kWh)	Total energy consumed per m^2 of dwelling (kWh)	Total energy bought (kWh)	Total energy bought per m^2 of dwelling (kWh)	Number of occupants
Department standard	4587	36	4587	36	0	21,786	172	10,639	84	10,639	84	1.7
LEH-A	3305	33	1605	16	2916	13,977	140	7188	72	5488	55	3
LEH-B	3495	35	1890	19	2497	26,044	263	10,140	102	9124	92	2
LEH-C	3978	40	1756	18	3257	16,614	166	8593	86	6371	64	1
LEH-D	3285	33	1604	16	2853	27,463	277	10,914	110	9233	93	1.5
ConA	4584	56	4584	56	0	32,776	400	13,688	167	13,688	167	2
ConB	2259	23	2259	23	0	55,864	576	17,777	183	17,777	183	2
ConC	1510	29	1510	29	0	14,827	285	5629	108	5629	108	1
ConD	5860	60	5860	60	0	32,776	338	14,964	154	14,964	154	3
ConE	2223	25	2223	25	0	30,491	346	10,693	122	10,693	122	2
ConF	2172	26	2172	26	0	24,618	290	9010	106	9010	106	2
ConG	3118	NA	3118	NA	0	14,008	NA	7009	NA	7009	NA	2

dwelling. Table 3 presents the monitored energy performance of the low-energy houses (LEH) and control houses (Con) in comparison to the design of the Department standards. The control households consumed less electricity (3104 kWh) when compared to the low-energy households (3516 kWh). When adjusted to include the solar generation, the low-energy households purchased 45% less electricity compared to the control households and 62% less electricity compared to the Department standards. The low-energy households were also found to consume 15% less gas when compared to the control households and 3% less gas than the Department standards.

Figure 4 presents the preceding electricity and gas data in a single graph for comparison. Overall, the low-energy houses used an average 12% less energy than the Department standards and 7% less energy than the control households. When solar generation is factored in, overall the low-energy houses purchased 29% less energy than the Department standards and 24% less energy than the Control households. This translated to the low-energy houses achieving 50% less environmental impact (CO_2^{-e}) compared to the Department standard and 40% less environmental impact compared to the control houses.

4.2 *Improved Thermal Performance in Summer*

While the direct performance of energy consumption and generation discussed above points to more sustainable housing, there was also the benefit of addressing energy efficiency with respect to the thermal performance of the dwellings, especially over the summer months. The low-energy houses were built to not require air conditioning.[1] Analysis of the summer time temperature data from the low-energy dwellings

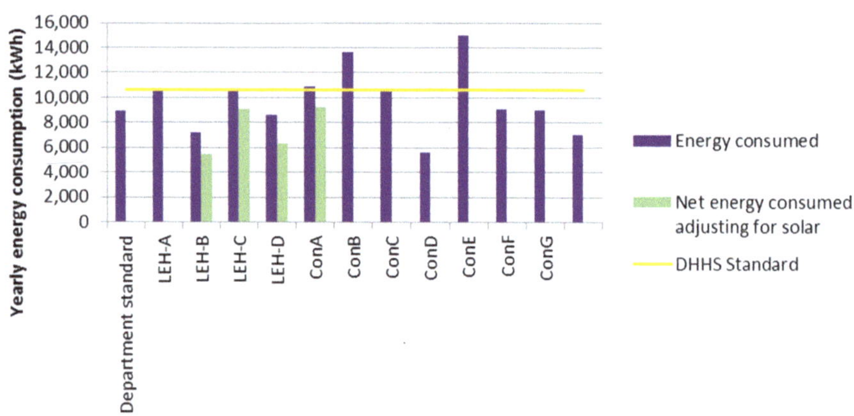

Fig. 4 Yearly energy consumption for each of the dwellings

[1]One of the low-energy houses installed air conditioning during the evaluation period due to perceived health issues (they believed they were more susceptible to pneumonia due to their age and

found that they had better thermal comfort compared to the control households and particularly during extreme weather even though the control houses had various types of air-conditioning systems. For example, during the summer period the average temperatures inside the low-energy houses (23.8 °C) and control houses (24.0 °C) were similar for the living area, but the low-energy houses had an average mean temperature of 1.2 °C lower for the bedrooms. However, the average maximum temperature in the living areas of the control houses was significantly higher (2.7 °C) compared with the low-energy houses.

The assessment of the adaptive comfort criteria against the European thermal adaptive comfort standard, BS EN 15251 using monitored temperature and humidity data shows that the low-energy houses were comfortable for 10% more of the time in summer for the living areas and 7% more of the time for the bedrooms compared with the control houses; this was all achieved without the use of additional air conditioning. The biggest benefit for thermal comfort was during extreme weather conditions such as heatwaves (with temperatures reaching upwards of 45 °C during the study period), when the low-energy houses were significantly cooler than the control houses which were using air conditioning, reflecting the improved design and thermal performance of the dwellings. Figure 5 shows that on the second day of a heat wave, the best low-energy house was 16.6 °C cooler compared to the worst control house (with air conditioning). At least one of the control households (without fixed air conditioning) found it too hot to stay in their dwelling during heatwaves and spoke about the negative impact of having to find other places to stay during such periods. He stated (ConB):

> One of my friends had a device and walked in here one day and it was like 51 degrees… if you're expecting a week of 40's…most of all my friends have got air conditioning so I normally sleep there…

This improved comfort particularly during the more extreme weather periods was something the low-energy residents spoke about during the interviews. For example, LEH-B stated:

> Well we both feel the heat pretty well but when it was 42 degrees outside, it only got to 29 in here…when it was 3 degrees below zero this was 15 degrees inside on that morning, that's without any heaters being on, 15 degrees. So that's good.

This improved thermal performance of the low-energy houses was noticed by the occupants in relation to self-reported health improvements. For example, one occupant used to get pneumonia regularly during winter in their previous accommodation, but had not had a case of it over the first three years in the low-energy houses; an outcome they relate directly to the improved, and consistent, thermal performance. Another occupant reported that they would get cramps in their legs when it got too cold, which made sleeping in winter difficult unless they were next to a heater. Again this had seen a dramatic improvement in the low-energy house due to the improved thermal comfort.

---

previous health issues). The monitored data before they installed the air conditioning suggested the dwelling remained comfortable over summer.

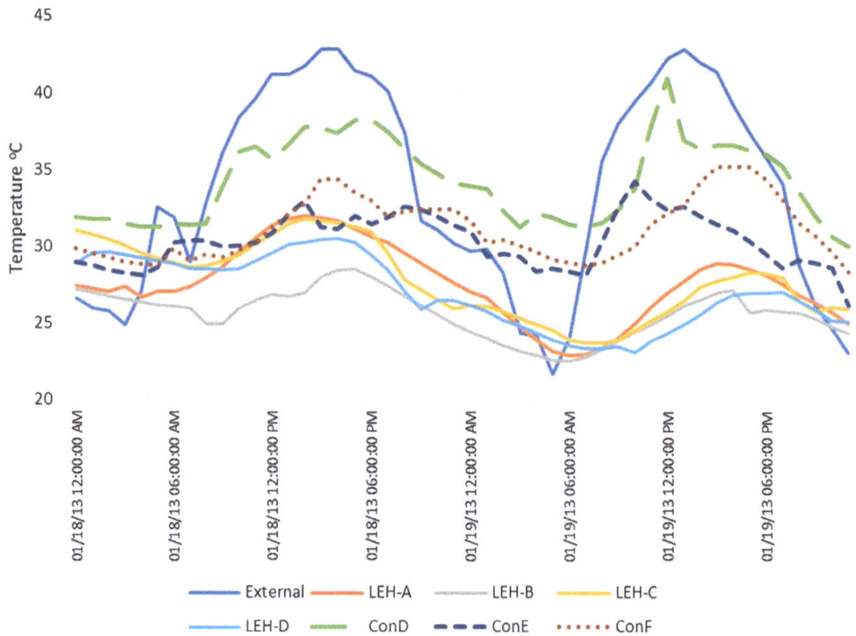

Fig. 5 Temperature in living rooms of monitored houses and external temperature for 18–19 January 2013

Despite the improved thermal performance of the low-energy houses, it was also evident to the researchers that the occupants in the low-energy houses were not always using the design and sustainability features as intended to help with the thermal comfort and overall sustainability of the dwellings. For example, one household was not using the celestial windows to help vent excess heat inside during summer as they believed the architect placed them on the wrong orientation, meaning the occupant believed they let heat in, rather than vented it out.

4.3 *Costs*

While the above energy, environmental and thermal comfort data all suggest a significant improvement from the low-energy houses, this must be all considered within the context of the cost to achieve such an outcome, especially for social housing providers who must balance the need for more overall housing with the need to improve outcomes from those in the housing. So, what were the costs for the project and is it feasible to be repeating?

The additional upfront cost for the low-energy houses was calculated to be $75,780 per dwelling (Table 4), which was found to be higher than for other similar sustainable

Table 4 Additional upfront costs of low-energy houses compared to standard department houses

Element	Cost per unit	Additional maintenance cost per year per unit ($)	Total cost for replacement across 40 years (includes inflation)
Building envelope	$55,322	$553	
Solar photovoltaic system	$9625	$96	$13,531
Rainwater tank plumbing and pump	$10,833	$23	$1673
Total	$75,780	$672	$15,204

Table 5 Summary of additional costs to the department

Element	Initial cost	Accumulated cost after 5 years	Accumulated cost after 40 years
Additional building envelope, solar photovoltaic, rainwater tank plumbed into house	$75,780	NA	NA
Additional maintenance	NA	$3570	$50,705
Additional solar photovoltaic and rainwater tank elements replacement	NA	NA	$15,204
Change to rent received	$0	$0	$0
Total additional cost to the department	$75,780	$79,350	$141,689

housing projects in Australia [24]. The majority of this cost was for the improved thermal performance of the building envelope. A maintenance costs and cost for technology replacement were also considered. At both a high- and low-energy price future, and for a discount rate of 3.5 or 7.0%, the low-energy houses do not achieve a positive payback within a traditional cost–benefit framing.[2]

While there are substantial costs to the Department over 40 years ($141,689 of which $75,780 is capital cost and $65,909 is additional maintenance and replacement of technologies—see Table 5), there are significant financial benefits to the households. The low-energy households saved an average of $1050 per household from the improved design. They also deliver significant contributions to environment, comfort and broader society benefits that are not costed in this study.

The low-energy households spoke about being better off financially in the low-energy dwellings. This was noticeable for them as it allowed them to do things they had been unable to do previously like buy presents for family members, go shopping without having to use lay-by and even go on a holiday. For example, LEH-D states:

[2]If assuming the Department received the solar feed-in tariff rates.

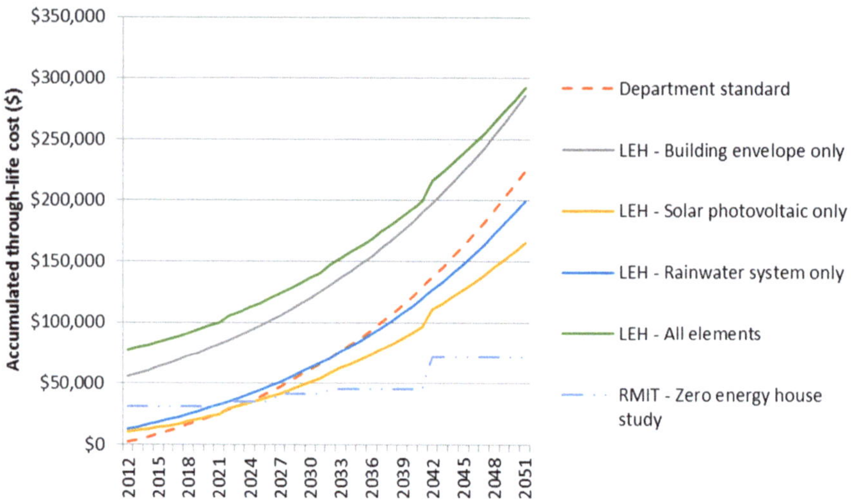

Fig. 6 Accumulated costs for various sustainability elements within the low-energy houses across time for a low-energy price future in comparison to a zero energy house study previously conducted at RMIT University [21]

> I do go clothes shopping on occasion now instead of thinking, "Oh God, I have to go and layby that.

If the additional upfront costs are broken down to their individual elements, there are some elements which are more financially viable than others (Fig. 6). What can clearly be seen is that for both a low and high energy price future, the solar photovoltaic system is the most cost-effective element, followed by the rainwater system plumbed into the house. The solar photovoltaic system has a payback period of 10–13 years, and the rainwater tank plumbed into the house has a payback period of 17–21 years. Only for the high energy price future does the building envelope only or the whole low-energy house achieve a payback (36 years) compared to the standard industry practice, and neither of these options achieve payback within the 40-year modelling against the Department standard. This indicates that it is more economically viable for the solar photovoltaic system and water elements than it is for the building envelope.

5 Discussion

The above analysis demonstrates that the low-energy houses performed significantly better than the control houses from an energy efficiency and thermal comfort perspective. This was not unexpected as the dwellings were specifically designed to be more energy efficient and to generate renewable energy to achieve a low-energy outcome. The question is how these low-energy houses impacts on social housing provision—an area which is often overlooked in the sustainable housing discussion.

5.1 Social Housing Providers

As stated in the introduction, the providers of social housing often face complex and sometimes competing objectives which must be balanced out. They are constantly balancing the need for more dwellings but also to improve occupant liveability, health, well-being and financial outcomes for low-income households, as well as contribute to broader government sustainability targets, an issue which is not just related to the Australian context [20]. Developing new housing, or purchasing existing housing, is a costly exercise and so due consideration must be given to ensure that the best value for money is achieved.

Anything that adds to these costs, as sustainability elements typically do, takes away the ability to obtain more housing. As the above study found, there was a significant cost for achieving the low-energy outcomes, which is arguably the biggest challenge the Department has to overcome if these housing are to be replicated. Broader research has found that achieving low-energy housing could be done for much less than what the Department spent [7] which means if the Department was to repeat this project, significant cost savings could be achieved. For example, the solar panels were found to be about twice the cost of average systems which was partly because of the regional location and constraint of choice in the marketplace.

An additional cost and challenge for social housing providers is not just the additional capital costs, but the ongoing maintenance costs for sustainability technologies such as solar photovoltaics and rainwater tanks. This included needing to factor in for replacement at end of life, and that there will inevitably be faults in these technologies/systems from time to time. While a regular maintenance program can be developed, without some type of remote monitoring of the systems it can be difficult to pick up on faults. One of the low-energy houses in this research spoke about how their solar photovoltaic system stopped working, and it was only when they received their utility bill which was higher than normal that they realised something was wrong. The additional challenges of maintenance and faults for social housing providers have been identified in other low-energy and social housing research [25].

Another challenge is that the Department does not have a mechanism for charging higher rent for their properties even if they have lower costs to live in. Currently, rent is set as a percentage of their total income—in Victoria where this case study is located that percentage is capped at 25% of gross income. Benefits from things such as lower energy bills or income generated through feed-in tariffs are not considered within that framework. To make sustainable housing more affordable for social housing providers, it may be that they need more innovative ways to recoup some of the sustainability costs. For example, perhaps the Department could have claimed half of the feed-in tariff, a situation which would have provided some additional money for the Department, but also ensured the tenant was better off as well.

However, social housing providers typically also have objectives around improving quality of life for tenants, such as through improving health outcomes or financial circumstances. In this regard, the low-energy houses were achieving beneficial outcomes. The occupants self-reported improved health outcomes due to improved thermal comfort. While not explored in detail in this project, reducing the number of trips to the doctors, or hospital stays, due to improved health outcomes resulting from improved thermal comfort has the ability to help reduce costs and congestion across the already stretched health care system.

5.2 Tenants

The challenges for the tenants related to how they used the low-energy houses. Despite being provided with a tour of the houses and having the various sustainability elements explained to them, and being provided a manual for the house, a number of tenants in the low-energy houses refused to use some of the sustainability features as designed. The previously mentioned exampled about the misuse of the clerestory windows to help vent heat in summer is a case in point. This raises questions about if such elements should be automated, or if the households should have control. Overall though the tenants were mostly following the directions on how to use the dwellings.

Another challenge for the tenants related to knowing when the low-energy houses were not performing as they should and how to address the problem. In one instance, a solar panel had failed but the householder did not become aware of this until their energy bill came in two months later and was significantly higher than it had been previously. It was only through contacting their energy provider that the failed solar panel was identified. While it might not be suitable for in-house monitoring for all sustainability elements to alert tenants to any issues, this might be something that the housing provider (in this case the Department) could monitor remotely.

Despite these challenges, there were significant benefits for the social housing tenants in the low-energy houses. For example, the improved energy performance and inclusion of solar photovoltaic systems resulted in the households being better off by $1050 a year in direct energy savings. This in turn meant that these low-energy households were more financially secure and had more money for spending on other areas of their life. One of the households had turned their financial situation around so significantly they no longer received CentreLink[3] payments; this was partially due to improved affordability of living in the low-energy house but also because their health had improved because of the better thermal comfort.

[3]Government welfare payment.

5.3 *Opportunities for Social Housing*

The benefits realised by the low-energy housing in this case study are in line with what other sustainable housing developments around the world are finding [8, 9, 22, 25, 29]. These benefits include improved environmental outcomes, lower purchased energy, improved thermal comfort, improved occupant health, improved occupant liveability and financial outcomes. As found in other research into low-income sustainable social housing around the world, there is not necessarily one policy or development outcome which will suit every social housing provider [26]. However, there are some key lessons which are applicable across different organisations.

The challenge now for the Department, and other such social housing providers, is to find a way to improve sustainability at a lower capital cost. One option would be to pull back on the thermal performance (e.g. back down to 8 stars NatHERS rating), but this would then mean the housing would not perform as well during extreme weather conditions and would likely require the inclusion of air conditioning which would add additional capitals costs for the systems and ongoing operating costs for the households. There have also been other building and technology innovation in the years since these low-energy houses were built, so there would likely be cost efficiencies that could be found, for example with the solar panels. There is also a need for the way that occupants use social housing to be better integrated into the design process to ensure that the housing performs as predicted [27].

5.4 *Limitations*

Due to the space limitations of this chapter, some elements from the above evaluation have not been explored in detail. Further details from the study, including additional data analysis (e.g. blower door tests) can be found in the main project report [24].

6 Conclusion

This chapter has explored the performance and outcomes of a low-energy social housing development in regional Victoria, Australia. The evidence finds that the houses performed extremely well in terms of energy efficiency. The houses also provided a number of benefits of the social housing tenants such as reducing energy bills, providing an energy rebate from the feed-in-tariff from the renewable energy generation and improving health and well-being outcomes. While there were many benefits, there were also several challenges both for the Department (e.g. high upfront costs) as well as the tenants (e.g. learning to use the houses as designed). It is clear though that the evidence supports aspirations by the Department to move beyond their current minimum housing standards for new construction. A combination approach, whereby the thermal performance of the dwelling is improved, in addition to including renewable energy generation, will address several goals of housing providers—namely improving quality of life, improving health outcomes, finances and environmental impacts.

Acknowledgements The construction, research and evaluation of this project were funded by the Director of Housing, Victoria and are reproduced with permission of the Director of Housing. The author thanks the research participants (householders and stakeholders) who generously gave their time to this project, Becky Sharpe and Daniel Voronoff from the Department of Health and Human Services, Ian Adams from Organica Engineering and acknowledges the wider RMIT research team involved in the project: Yolande Strengers, Cecily Maller, Larissa Nicholls, Ian Ridley, Ralph Horne and Shae Hunter.

References

1. ABS (2015) 6523.0—household income and wealth, Australia, 2013–14. Australian Bureau of Statistics, Canberra
2. ACOSS (2013) Energy efficiency & people on low incomes. Australian Council of Social Service, Sydney
3. AIHW (2017) Housing assistance in Australia 2017. Retrieved 18 Nov. 2017, from https://www.aihw.gov.au/reports/web/web-189/housing-assistance-in-australia-2017/contents/social-housing-tenants-1
4. Australian Energy Regulator (2016) Annual report on the performance of the retail energy market 2015–16. Australian Energy Regulator, Melbourne
5. Australian Energy Regulator (2017) State of the energy market May 2017. Australian Energy Regulator, Melbourne
6. BCEC (2016) Energy poverty in Western Australia. A comparative analysis of drivers and effects. Bankwest Curtain Economics Centre, Perth
7. Berry S (2014) The technical and economic feasibility of applying a net zero carbon standard for new housing. University of South Australia
8. Berry S, Whaley D, Davidson K, Saman W (2014a) Do the numbers stack up? Lessons from a zero carbon housing estate. Renew Energy 67:80–89
9. Berry S, Whaley D, Davidson K, Saman W (2014b) Near zero energy homes—what do users think? Energy Policy 73:127–137
10. Boardman B (2012) Fuel poverty. In: Smith SJ (ed) International encyclopedia of housing and home. Elsevier, San Diego, pp 221–225
11. Chester L (2013) The impacts and consequences for lowincome Australian households of rising energy prices. University of Sydney, Sydney
12. DBEIS (2017) Annual domestic energy bills. Average annual domestic electricity bills for UK Countries (QEP 2.2.2). E. I. S. Department for Business. London, Department for Business, Energy & Industrial Strategy
13. DHHS (2017) Social housing. Retrieved 18 Nov. 2017, from http://housing.vic.gov.au/social-housing
14. EIA (2017) Short-term energy and summer fuels outlook. U.S.E. I. Administration. Washington, U.S. Energy Information Administration
15. Essential Services Commission (2016) Victorian energy market report 2015–16. Essential Services Commission, Melbourne
16. Farbotko C, Waitt G (2011) Residential air-conditioning and climate change: voices of the vulnerable. Health Promot J Austr 22(Spec No: S13–S16)

17. Hoppe T (2012) Adoption of innovative energy systems in social housing: lessons from eight large-scale renovation projects in The Netherlands. Energy Policy 51(Supplement C):791–801
18. Jacobs Australia Pty Limited (2016) Retail electricity price history and projections—public. Jacobs Australia Pty Limited, Melbourne
19. Langevin J, Gurian PL, Wen J (2013) Reducing energy consumption in low income public housing: interviewing residents about energy behaviors. Appl Energy 102(Supplement C):1358–1370
20. Marzouk M, Azab S (2017) Analyzing sustainability in low-income housing projects using system dynamics. Energy Build 134(Supplement C):143–153
21. Moore T (2012) Facilitating a transition to zero emission new housing in Australia: costs, benefits and direction for policy. RMIT University
22. Moore T, de Haan F, Horne R, Gleeson B (2018) Urban Sustainability Transitions. Australian Cases—International Perspectives. Springer, Singapore
23. Moore T, Ridley I, Strengers Y, Maller C, Horne R (2016) Dwelling performance and adaptive summer comfort in low-income Australian households. Build Res Inf 1–14
24. Moore T, Strengers Y, Maller C, Ridley I, Nicholls L, Horne R (2016) Horsham catalyst research and evaluation. Final Report. RMIT University, Melbourne
25. Pretlove S, Kade S (2016) Post occupancy evaluation of social housing designed and built to code for sustainable homes levels 3, 4 and 5. Energy Build 110(Supplement C):120–134
26. Roders M, Straub (2015) A assessment of the likelihood of implementation strategies for climate change adaptation measures in Dutch social housing. Build Environ 83(Supplement C):168–176
27. Santangelo A, Tondelli S (2017) Occupant behaviour and building renovation of the social housing stock: current and future challenges. Energy Build. 45(Supplement C):276–283
28. Smith A (2006) Governance lessons from Green Niches: the case of eco-housing. In: Murphy J (ed) Governing technology for sustainability. Earthscan, London, pp 89–109
29. US EPA (2011) Energy efficiency in affordable housing. A guide to developing and implementing greenhouse gas reduction programs. Local Government Climate And Energy Strategy Series. Washington, United State Environmental Protection Agency (2011)
30. Victorian State Government (2007) K2 apartments technical report. Retrieved 05 Aug. 2017, from http://www.dhs.vic.gov.au/__data/assets/pdf_file/0003/560964/k2_apartments_technical_report.pdf

Dr. Trivess Moore B.Sc. (Env.), Ph.D. is a research fellow with expertise in energy efficiency, renewable energies, the built environment and transitioning to a low carbon future.

Indoor Environmental Quality of Preparatory to Year 12 (P-12) Educational Facilities in Australia: Challenges and Prospects

Mary Myla Andamon and Jin Woo

Abstract Climate change is leading to increased frequency, intensity and duration of heatwaves not only in Australia but globally. Children are among those who are most physically vulnerable to the changing climate. Schools buildings and facilities are critical infrastructure which are at risk of the adverse impacts of extreme weather conditions, particularly to the schools' indoor environments. This chapter reviews the diverse policies on cooling and ventilation in educational facilities across Australia and brings together a multidisciplinary appraisal which can provide starting points for designers, building scientists and policy makers on:

- Impact of building energy efficiency measures on the thermal comfort, IAQ and ventilation of educational facilities.
- Health, educational outcomes and economic impacts of thermal comfort, IAQ and ventilation within educational facilities.
- Australian and best practice international policies, standards and practices applicable to the thermal environment, IAQ and ventilation within P-12 educational facilities.

1 Introduction

While other chapters in this book look at building energy and sustainability performance, this chapter explores indoor environment performance of school buildings by looking at national and international standards, design guidelines and policies on indoor environmental quality (IEQ) for Australian Preparatory to Year 12 (P-12) educational facilities.[1] This examines the relationship between the IEQ parameters

[1]See Chapter "University Buildings: the Push and Pull for Sustainability" for a discussion on sustainability in university buildings in Australia.

M. M. Andamon (✉) · J. Woo
Sustainable Building Innovation Laboratory, School of Property Construction and Project Management, RMIT University, Melbourne, VIC, Australia
e-mail: mary.andamon@rmit.edu.au

P. Rajagopalan et al. (eds.), *Energy Performance in the Australian Built Environment*, Green Energy and Technology, https://doi.org/10.1007/978-981-10-7880-4_8

of thermal comfort, indoor air quality and ventilation and educational outcomes for the P-12 group. Knowledge gaps exist in this area with limited research undertaken in Australia to establish potential benefits of indoor environmental quality improvements in schools. This chapter establishes the need for a study grounded on addressing the absence of clear documentation on the state of indoor environments in educational facilities in Australia backed by measurements and surveys of temperature, comfort conditions, indoor quality and the relationship between these aspects of indoor environments and academic performance of students.

2 Indoor Environmental Quality in Educational Facilities

Extremely hot weather conditions are becoming more commonplace and severe in Australia[2] [12]. It is projected that by 2070, Adelaide, Sydney and Melbourne can expect to experience at least twice as many days with extreme temperatures, while residents of Darwin could find 35 °C days occurring for up to two-thirds of the year [61, 72]. In Melbourne, the long-term annual average number of days above 35 °C is 10 but rose to 13 during the decade 2000–2009 [11]. The number of hot days is likely to increase to 15–26 days by 2070 [13]. The increasing number of hot days will impact on public health, mortality rates, energy demand and economy of Australia [43, 61, 76].

With significant increases in the frequency of extreme temperatures in the country [62], public concern about the adverse effects of school indoor environments in Australia has increased in recent years [18, 70, 71]. School buildings, as with much of the country's critical infrastructure and facilities, are vulnerable to poor performance during extreme weather. Poor indoor air quality (IAQ) and thermal conditions are known to decrease productivity and cause dissatisfaction for building occupants [44, 79]. However, much of the research on indoor environments focuses on adult workers in offices [22]. There is limited information on the relationship between indoor environments and the learning performance and behaviour of children in school buildings [47, 75].

In 2016, there were 344,726 children aged four or five years enrolled in a preschool (early childhood development) program in Australia [2], where 43% attend preschools and 51% long day care (LDC) centres. These preschoolers attend for 15 hours or more per week or up to 645 hours in a school building annually. Students in Years 1 to12 receive at least 25 hours of instruction per week [68] or spend up to 1075 hours indoors in school buildings annually. Australian students will spend up to 12,900 hours of their lives in school buildings from Preschool to Year 12—which

[2]Chapter "Urban Climates in the Transformation of Australian Cities" provides a discussion on the changing climate trends in Australia.

would be up to 25% of their waking lives to the completion of their schooling [17]. With the number of hours spent in classrooms, the conditions of indoor environmental quality factors in school buildings and their impact on children's health,[3] well-being, comfort and learning ability remain a subject area of concern [25, 74].

3 Indoor Environmental Quality (IEQ)

The terms *indoor environmental quality* (*IEQ*) and *indoor air quality* (*IAQ*) have different meanings. Both terms refer to environmental qualities within a building, and they are used especially in relation to the health and comfort of building occupants. However, IEQ is a broader concept that includes IAQ as one of the indoor quality elements [8], including (1) thermal comfort, (2) indoor air quality and ventilation, (3) lighting levels and (4) acoustics and noise. However, this chapter focuses on thermal comfort, IAQ and ventilation as they are the dominant factors to achieve an overall comfort (or a high level of IEQ) among the indoor elements. Acoustics and noise, lighting quality and levels (including daylighting), odour quality (olfactory) and visual perception are not included in this review.

Much research has been conducted on thermal comfort, IAQ and ventilation [20, 27, 58]. Thermal comfort and IAQ affects an occupant's well-being, health, perception of IEQ and productive performance [8, 21]. This general statement applies to children's health and learning achievements in school buildings. An extensive body of knowledge on thermal comfort and IAQ has informed the development of international standards and guides for the design and management of the indoor environments of buildings. Australian standards, guides and codes of practice typically follow international developments.

3.1 Thermal Comfort

The primary environmental factors that determine thermal comfort are [37] air temperature, radiant temperature, humidity and airspeed. The primary personal parameters are clothing and activity level. Discomfort can be minimised in various ways. In a warm or hot environment, the amount of clothing or level of physical activity can be reduced, or an environment that is more conducive to increased heat loss can be created (see also Chapter 'Thermal Environments in the Construction Industry: A Critical Review of Heat Stress Assessment and Control Strategies'). Conversely, in a cool or cold environment, the responses could include increased clothing, increased activity, or seeking or creating an environment that is warmer [10]. While thermal

[3] Although this chapter mainly focuses on the student requirements for indoor environmental quality of school facilities, the authors acknowledge that the same indoor conditions would affect the teaching and administration staff.

comfort is essentially a subjective response and may be influenced by contextual and cultural factors (i.e. psychological adaptation) [29], a person's sense of thermal comfort is primarily a result of the body's heat exchange with the environment.

The development of international thermal comfort standards has largely been based on studies of healthy, fit and sedentary office workers across a range of climate zones: temperate, hot and humid, cold and hot-arid [30]. However, research indicates that adult-based thermal comfort standards are not directly applicable to children in school buildings [48, 65, 66]. Teli et al. [65] suggest this may be due to children's higher metabolic rate (per kg of body mass); their limited ability to adapt to the indoor thermal environment by controlling heaters, coolers, windows, blinds, etc., in the classroom, or by changing their clothing; and/or possibly their strong relationship with the outdoor climate since their daytime schedule includes outdoor activities and play, unlike office workers.

3.2 Indoor Air Quality (IAQ) and Ventilation

Indoor air quality directly impacts occupant health, comfort and work performance. People in buildings frequently report building-related health symptoms and sometimes develop building-related illnesses. Research has shown these health and comfort effects are associated with the characteristics of buildings, HVAC systems and the indoor environment [20, 63]. The *American Society for Heating, Refrigerating and Air-Conditioning Engineers* (ASHRAE), *International Organization for Standardization* (ISO) and the *European Committee for Standardization* (CEN) have developed standards and guides for indoor air quality and ventilation and the criteria and recommendations in the international standards have been adopted as normative references by Australian national standards, guidelines and codes of practice [23, 24, 59, 60, 77]. In contrast to the specifications for thermal environments, it has not been possible to agree on a method for specifying the level of indoor air quality in buildings. Instead, required ventilation rates are specified for different types of space and occupancy [54].

4 Design Guides and Best Practice Standards

Design guides and standards on the approaches to cooling and heating, thermal comfort and ventilation in educational facilities in Australia follow *ASHRAE Standards 55* [9] and *62.1* [7] in North America and *ISO 7730* [42], *CIBSE Guide A* [19] and *CR 1752* [14] in Europe. Although the methodologies underpinning these standards differ, a deterministic stimulus-response approach based on laboratory methods is used in *EN ISO Standard 7730* while a holistic person–environment systems approach based on field research is used in *ANSI/ASHRAE Standard 55* [27]—both evaluate the general thermal state of the body based on a heat balance analysis. These thermal

Table 1 Recommended indoor temperatures and relative humidities for general use (*Sources* ASHRAE Standard 55 [9], EN ISO 7730 [42] and CIBSE Guide A [19])

Standard	Thermal comfort criteria
ASHRAE Standard 55[a]	Winter: Temp 19–26 °C Summer: Temp 23.5–28 °C (graphical method)
ISO 7730[a]	Winter: Temp 20–24 °C; RH 30–70% Summer: Temp 23–26 °C; RH 30–70%
CIBSE Guide A[a]	Temp 22–24 °C, RH 30–60%

[a] Light, mainly sedentary activity

Table 2 Recommended indoor temperatures for Australia and Victoria (*Sources* Comcare Australia and Community and Public Sector Union [24], WorkSafe Victoria [77, 78])

Standards, guides and codes	Thermal comfort criteria
Comcare Australia and Community and Public Sector Union [24] [a]	Summer: Temp 23–26 °C
WorkSafe Victoria [77][a]	All year: Temp 20–26 °C; Airspeed 0.1–0.2 m/s

[a]Mainly sedentary work

comfort standards prescribe numeric and descriptive criteria for comfort primarily for mechanically conditioned spaces (19.7–26.7 °C, 20–60% RH). For free-running (non-conditioned) buildings and during warm weather, 25 °C is an acceptable indoor design temperature.

4.1 Criteria for Thermal Environments in Educational Facilities

International standards and guides specify environmental conditions for acceptable thermal comfort in terms of indoor temperatures and humidities (Table 1) or predicted mean votes (PMV) [42]. These have been adopted into guides and codes for offices in Australia (Table 2). The Victorian Trades Hall Council (VTHC) endorses these recommendations and circulates them as part of occupational health and safety (OHS) information for the education sector in Victoria [69]. Environmental conditions and the level of thermal comfort expected in a building depend on the type of building and its occupants. Conditions for educational facilities are recommended by ASHRAE [6] (see Table 3).

Table 3 ASHRAE recommended indoor temperatures and humidities for educational facilities (*Source* ASHRAE Handbook—HVAC Applications [6, Tables 1 and 6, pp. 7.1–7.4])

Category	Occupancy	Humidity criteria[a]	Temperature (°C)[a]	
			Winter	Summer
Preschools	Infant, toddler, and preschool classrooms	30% RH	20.3–24.2	23.8–26.7
		40% RH	20.0–23.9	23.1–26.7
		50% RH	20.3–23.6	22.8–26.1
		60% RH	19.7–23.3	22.8–25.8
	Administrative areas, offices, lobbies and kitchens	RH 30–60%	20.3–23.3	23.3–25.8
K-12 schools	Classrooms, laboratories, libraries, auditoriums and offices	30% RH	20.3–24.2	23.3–26.7
		40% RH	20.0–23.9	23.1–26.7
		50% RH	20.3–23.6	22.8–26.1
		60% RH	19.7–23.3	22.8–25.8

[a]Based on EPA [35] and ASHRAE Standard 55 [4] for people wearing typical summer and winter clothing, at mainly sedentary activity

4.2 *Criteria for Indoor Air Quality (IAQ) and Ventilation in Educational Facilities*

Standards and guides prescribe minimum ventilation rates as a means of achieving acceptable indoor air quality [39]. The relationship between indoor air quality, in terms of CO_2 concentration levels and ventilation rates is shown in Table 4. Standards and reference guides include *ASHRAE Standard 62: Ventilation for Acceptable Indoor Air Quality* [7], European Standard *EN 13779: Ventilation for non-residential buildings—Performance requirements for ventilation and room-conditioning systems* [15] and the technical report *CR 1752-1998: Ventilation of buildings—Design criteria for the indoor environment* [14].

The prescriptive method of *ASHRAE Standard 62.1* adds the minimum ventilation rate per person to the minimum ventilation rate per square metre of floor area. The person-related ventilation accounts for pollution from people and the ventilation rate based on floor area accounts for emissions from building materials, furnishings, HVAC system, etc. [54]. A similar approach is used in *CR 1752*. However, only person-related ventilation is required if it is assumed the building does not emit pollution [54]. Ventilation rates for education facilities prescribed by *ASHRAE Standard* 62.1 and CR 1752 are compared in Table 5.

Table 4 Classification of indoor air quality (IDA) according to EN 13779 (*Source* Olesen [54, p. 22])

Category	Description of indoor air quality	Classification parameter	Ventilation rate (outdoor air)	
		CO_2 level above outdoors (ppm)	Non-smoking (l/s per person)	Smoking (l/s per person)
IDA 1	High	≤400	>15	>30
IDA 2	Medium	400–600	10–15	20–30
IDA 3	Acceptable	600–1000	6–10	12–20
IDA 4	Low	>1000	<6	<12

Categories of indoor air quality as specified in EN 13779 [15, p. 19]

Ventilation requirements for educational facilities in the Australian Standard *AS 1668.2-2012: The use of ventilation and air conditioning in buildings—Mechanical ventilation in buildings* [59] align with *CR 1752* (see Table 6).

4.3 Interactions Between Thermal Comfort and Indoor Air Quality (IAQ)

IAQ directly impacts occupant health, comfort and work performance. Providing superior IAQ can improve health, work performance and school performance, as well as reduce health care costs, and consequently be a source of substantial economic benefits [16, 79].

Temperature is recognised as a key factor for human comfort. However, less attention may have been given to the importance of humidity [50, 67]. Recent studies on the direct impact of temperature and humidity on human perception of IAQ found that acceptability of air decreased with increasing air temperature and humidity levels [36, 57]. Changing the air temperature or humidity of the indoor environment may change IAQ by significantly affecting the emission source strength of materials in a space and the perception itself of air due to change in chemical composition [41]. Maintaining dry and cool indoors as opposed to humid and warm may improve both the perceived air quality and ventilation requirement. For example, in a study by Fang et al. [36], reducing the ventilation rate from 10 to 3.5 L/s per person in an office space can be compensated for by reducing the air temperature and humidity from 23 °C and 50% RH to 20 °C and 40% RH, so as to avoid deteriorating perceived air quality. IAQ is typically addressed through compliance with minimum ventilation requirements in building regulations, which are based on industry consensus standards such as the *ANSI/ASHRAE Standard 62.1* [5]. The increased interest and attention on the impact of IAQ in buildings saw the publication of two guides which present best practices for design, construction and commissioning of buildings and provide information and guidance on IAQ-related issues in schools [3, 8, 35]:

Table 5 Recommended ventilation rates in educational facilities according to ASHRAE 62.1 and CR 1752 (*Source* Olesen [54, Appendix 2, p. 26])

Room type	Occupancy (person/m^2)	Category of indoor air quality	Minimum outdoor airflow rate (l/s per person)		Additional ventilation for building emissions (l/s per m^2)			Total (l/s per m^2)	
			ASHRAE	CR	CR low-polluting building	CR *Not* low-polluting	ASHRAE (R_a)	CR Low-polluting	ASHRAE
Classroom	0.5	A	3.8	10	1.0	2.0	0.3	6	2.2
		B	3.8	7	0.7	1.4	0.3	4.2	2.2
		C	3.8	4	0.4	0.8	0.3	2.4	2.2
Kindergarten	0.5	A	5.0	12	1.0	2.0	0.9	7.1	3.4
		B	5.0	8.4	0.7	1.4	0.9	4.9	3.4
		C	5.0	4.8	0.4	0.8	0.9	2.8	3.4

Category refers to the quality level of indoor air and corresponds to the percentage of dissatisfied: Class A—15% dissatisfied, Class B—20% dissatisfied, Class C—30% dissatisfied

Table 6 Minimum ventilation rates for educational facilities according to AS 1668.2 (*Source* AS 1668.2-2012 [59, Appendix A, p. 61])

Occupancy type	Net floor area per person[a] (m^2)	Minimum outdoor airflow rate (l/s per person)
Classrooms serving persons up to 16 years of age	2	12.0
Classrooms serving persons over 16 years of age	2	10.0
Laboratories	3.5	10.0

[a]Applies when number of occupants is not known. Where the occupancy is not indicated, the actual occupancy shall be determined during the design of the room

5 Policies and Protocols for Educational Facilities in Australia

No unified set of policies and guidelines exist for air conditioning (heating, cooling and ventilation) of Preparatory to Year 12 (P-12) educational facilities in Australia. While international standards such as *ASHRAE Standards 55 and 62.1* and guides such *CIBSE Guide A* are referenced, each state has its own unique set of policies and guidelines. These are summarised in Table 7. Climate zones form the basis of the cooling policy of Victoria [32, Sect. 5.8, p. 89]. Schools are cooled if they are located in NatHERS[4] Zones 20 or 27. These zones are hot and dry during the summer, with the mean maximum air temperature exceeding 30 °C during January. This is similar to New South Wales' policy of providing cooling for schools in locations where the mean maximum air temperature during January exceeds 33 °C, although New South Wales also accounts for the effect of building design on cooling demand by allowing 'hot spots' classrooms to be cooled when the mean maximum temperature is 30–33 °C. The Relative Strain Index (RSI) used in Western Australia is also a location-based cooling policy. The Cooler Schools zones in Queensland are likewise based on the climate map of the state. In contrast, the cooling policy of South Australia does not specify geographical locations. The state's policy takes a performance-based approach to cooling by specifying indoor air temperature requirements. Adopting a performance-based approach to cooling in educational facilities would objectively meet the appropriate requirements of young students.

Summarising the cooling policies of the five (5) Australian states, without critique, the climate conditions of NatHERS Zones 20 and 27 currently adopted by the Victorian Department of Education and Training and the Victorian School Building Authority [32] loosely corresponds to those in the geographical locations of Zones 4 and 6 of the National Construction Code—Building Code of Australia (NCC-

[4]Nationwide House Energy Rating Scheme (NatHERS) is a star rating system in Australia that rates the energy efficiency of a home, www.nathers.gov.au. See Chapter "The Built Environment and Energy Efficiency in Australia: Current State of Play and Where to Next" for a discussion on the energy efficiency regulatory frameworks in Australia.

Table 7 Australian policies and protocols on air conditioning (heating, cooling and ventilation) in schools

Policies and protocols
Victoria (VIC) The Building Quality Standards Handbook [32] of the Department of Education and Training and the Victorian School Building Authority provide guidance on the policy requirements on thermal comfort, cooling, heating and ventilation: • Cooling systems are provided to schools on the basis of their location within the Nationwide House Energy Rating Scheme (NatHERS) – Zones 20 and 27. All schools in these areas receive full air conditioning to their entitled spaces under the space and area guidelines. The remaining schools are not provided with cooling systems except in limited number of circumstances, e.g. information technology server rooms [32, Sect. 7.8.2, p. 97] • Air conditioning is provided to all special development schools regardless of location and all relocatable buildings. [32, Sect. 7.8.2, p. 97] • Thermostat setting for cooling should not be lower than 26 °C, for heating should not be higher than 18 °C [32, Sect. 7.8.4, pp. 101-102] • No cooling system should be installed until an energy target has been established and the performance of the proposed system compared with that target, and revised if necessary. [32, Sect. 7.8.2, p. 98] • Ventilation conforms to the BCA requirement on a minimum area proportional to the occupied room floor area. Fixed or opening devices must be 5% of the total floor area [32, Sect. 7.7]
New South Wales (NSW) The Air Cooling Policy of the NSW Department Education and Communities [52] ensures that: • Schools with a mean maximum January temperature of 33 °C or above are provided with air cooling to all habitable spaces • Schools with a mean maximum January temperature between 30 and 33 °C are eligible to apply for air cooling of 'hot spots' classrooms • Air cooling is provided to all demountable classrooms and libraries in NSW public schools • The department is developing a Thermal Comfort and Resource Efficiency Framework that aims to maximise the performance of existing buildings through passive design measures (such as roof insulation and sunshades) complemented where necessary by mechanical systems to meet extreme heating and cooling requirements. [53]
South Australia (SA) The Air Conditioning Protocol (SV001) of the SA Department of Education and Child Services [31] provides guidance on the policy requirements for the provision of air conditioning in public schools: • Learning areas in schools and children's centres shall have heating and cooling equipment capable of maintaining temperatures within the range of 20–26 °C when the outside temperature is between 6.5 and 37 °C (for Adelaide). When the outside temperatures fall outside these 'design temperatures', then room temperatures may be below 20 °C in winter and above 26 °C in summer • General learning areas, learning support areas and administration areas in schools and children's centres are to have temperatures maintained within the range of 20–26 °C on a Design Day as per comfort conditions detailed by ASHRAE Standard 55. This identified comfort conditions are being where there is a dissatisfaction rate of less than 10% • For the Adelaide metropolitan area, the 'design temperatures' are 6.5 °C for winter and 37 °C for summer. For design temperatures for other parts of South Australia refer to Australian Institute of Refrigeration Air Conditioning and Heating (AIRAH) Application Manual DA9-Air Conditioning Load Estimate • The minimum ventilation rate in learning areas shall be 10 litres per second per student and assuming a maximum capacity of general learning area (GLA) classrooms of 30 students • Students are dismissed at 12.30 pm on days when the forecast maximum is 38 °C or higher, or up to one hour before normal dismissal time when the estimated maximum temperature is to be at least 36 °C

(continued)

Table 7 (continued)

Policies and protocols
Western Australia (WA) The WA Department of Education uses the redefined 25-day Relative Strain Index (RSI) line and extended 20 day RSI as the boundary for the 'Air Cooling Zone' [70]: • All new schools within the 20 day RSI boundary will be provided with air cooling/air conditioning to the extent required • All existing schools within the 20 day RSI boundary will be eligible to have air cooling/air conditioning into classrooms and offices where air cooling has not previously been installed • The 20 day RSI line is to be considered a general guide (rather than a fixed demarcation line) that allows schools east and north of the line to qualify for air cooling/air conditioning
Queensland (QLD) The Queensland Government introduced the 'Cool Schools' program in 1996 and 'Cooler Schools' in 1998 [55]. These programs: • Assist both state and private schools in North Queensland to assess their building stocks and provide some cooling strategies, where needed • Recommend cooling classrooms only when the indoor temperature exceeds 27 °C • Implement passive cooling techniques/strategies, such as replacing sliding windows with louvers, installation of insulation. • Have no temperature limit for dismissing students [18] The Design Standards for DETE Facilities [34] states: • Schools located in the Cooler Schools zones are provided with air-conditioning systems. It is the intention that air conditioning is used only during the hot summer periods, and natural ventilation is used for the remainder of the year • The provision of natural ventilation in rooms that are not air-conditioned: Rooms designed for use by more than 15 occupants shall have external windows/doors/skylights with a minimum open-able area of 10% of floor area. Open-able windows and doors to be located on opposite sides of a room where possible [34, Sect. 3.2.1, p. 14] • Air supply rates for kindergarten and prep spaces should meet the higher rate of 12 l/s per person specified in AS 1668.2 applies [34, Sect. 3.5.1, p. 32] The Schools Standard Air Conditioning Specification [33] specifies the following air-conditioning design and performance parameters (Sect. 2.3.1, p. 31): • Summer: 26 °C ± 1 K DB, 55% RH (not controlled) • Winter: 21 °C ± 1 K DB, 55% RH (not controlled)

BCA) Climate Zones [1], characterised by hot, dry zone with average January maximum temperature of above 30 °C. New South Wales' policy of air cooling for schools with mean maximum January temperatures of 33 °C or above and 'hot spots' classrooms for 30–33 °C, likewise approximately correspond to Zone 4 locations in the NCC-BCA Climate Zones [1]. With reference to Western Australia's cooling policy which follows the RSI index, the geographical locations of the 20- and 25-day 0.3 RSI line correspond to those within Zones 4, 3 and 1 of the NCC-BCA Climate Zones. The cooling policies in South Australia and Queensland do not specify geographical locations but have outlined the conditions of occupied school spaces which require the provision of air conditioning.

6 Indoor Environmental Quality (IEQ) and Educational Outcomes

Reviews on research findings on the relationships of US school facility conditions to student achievement and behaviour indicate that the following thermal comfort factors correlate with positive educational outcomes (McGuffey 1982 cited in [75, 82]:

- A significant relationship between the thermal environment of a classroom and student achievement and behaviour.
- There was a consistent pattern of higher achievement in air-conditioned schools.
- Achievement was greater in facilities that allowed for individual preferences for heat.
- Excessive temperatures caused stress in students.
- Solar heating through glass is a major contributor to overheated classrooms.

However, limited data and inadequate clear documentation are available on the effects of poor indoor environments, particularly of thermal effects and indoor environmental quality on the performance of schoolwork by students. Because little research has been reported on these relationships for children in schools, much of the information have assumed that influences of indoor settings on adults have relevance to the influences of school environments on children [79, 81].

6.1 *Recent IEQ Research*

A recently completed study in Europe is the Schools Indoor Pollution and Health: Observatory Network in Europe (SINPHONIE) project [40]. It was the first pilot project to monitor the school environments in 25 European countries in parallel. The SINPHONIE project established a scientific/technical network to act at the EU level with the long-term perspective of improving air quality in schools and kindergartens to reduce the risk and burden or respiratory diseases among children and teachers due to outdoor and indoor air pollution. The SINPHONIE results were mainly on the causal relationships between exposure and health effects. However, the final report outlined that the most striking results overall are those that underline the relevance of IAQ in schools as a societal problem with clear impacts on the health, quality of life and learning performance of European schoolchildren [25].

The thermal comfort studies of New South Wales' schools undertaken by the NSW Department of Education and Communities in collaboration with the University of Sydney's Faculty of Architecture, Design and Planning Indoor Environmental Quality (IEQ) Laboratory and NSW Public Works [51] show that about 22.5 °C operative temperature was found to be the students' neutral and preferred indoor temperature, which is generally cooler than expected for adults under the same thermal environmental conditions [28]. Despite the lower-than-expected thermal neutrality, the school children demonstrated considerable adaptability to indoor temperature variations, equating to approximately 4 °C operative temperature. This comfort study was part of the department's *Thermal Comfort Framework* program and primarily aimed to maximise the performance of existing public school buildings through passive design measures complemented where necessary by mechanical systems to meet extreme heating and cooling requirements [52, 53]. It was anticipated that the benefits of this approach to thermal comfort, among others, will increase the number of learning spaces that provide comfortable learning environments and reduction in electricity consumption. However, based on available information, although the study includes environmental monitoring of the schools, the effects on school performance and educational outcomes were not assessed.

In Victoria, a study on a selection of schools primarily in Melbourne confirms that schools reflect poor air quality, ventilation and comfort control [45, 46] and the findings indicate that CO_2 concentration levels (>2,700 ppm), ventilation rates and air temperatures in classrooms during winter are non-compliant with the standards.

6.2 Student Performance

Mendell and Heath [47] carried out a review of research into the factors that influence student performance. The review highlighted that the direct association of thermal conditions of higher temperature [56] and lower relative humidity (Green 1974 cited in [47] on performance or attendance are significant in the decrease in beneficial outcomes. Schoer and Shaffran [56] found a general advantage for performance tests in the cooled environments (22.5 °C), with a consistent tendency for greater, statistically significant benefits for more complex performance tests. Mendell and Heath [47] were not able to assess the relationships between HVAC thermal control systems and performance or attendance due to unavailability of findings from studies. Although the findings of McNall and Nevins (1967 cited in [47]) were characterised as '*non-persuasive*' (p. 35), the study's comparison between one air-conditioned school and several non-air-conditioned schools in Florida (USA) found trends in favour of higher academic achievement in the air-conditioned school. However, these studies are decades old and updated research is required to validate these findings.

The most recent field study carried out in school classrooms was conducted by Wyon and Wargocki [80] in Denmark. The study sought to determine whether classroom air quality affects schoolwork. These field experiments show that reducing moderately high classroom air temperatures in late summer from the region of 25 to 20 °C by providing sufficient cooling and increasing effective outdoor supply rate from 5 l/s per person to 10 l/s per person, improved the performance of numerical and language-based tasks resembling schoolwork [80].

While inadequate ventilation is often suspected to be an important condition leading to reported health symptoms [38, 63, 64], ventilation rates have rarely been measured in schools [26, 47]. *ASHRAE Standard 62.1* recommends a minimum ventilation rate of 6.7–8.6 l/s per person for educational facilities. In a 1984 study of

11 randomly selected Danish schools, the reported ventilation measurements ranged from 1.8 to 15.4 l/s per person with an average of 6.4 l/s per person. European standards—*CR1752* and *CIBSE Guide A* recommend a minimum ventilation rate of 10–12 l/s per person. A more recent study on the ventilation rates of four naturally ventilated secondary schools in the UK was conducted during the heating season 2005–2006 [49] and found measurements that ranged from 3.9 to 10.5 l/s per person.

7 Conclusions and Research Imperatives for Educational Facilities

The primary objectives of this chapter were to review IEQ design guides, standards and policies for Australian P-12 educational facilities, survey the literature related to the relationship between educational outcomes and thermal environment and indoor air quality for the P-12 group and identify findings of applicable IEQ research.

7.1 Standards and Design Guides on Thermal Comfort, Indoor Air Quality (IAQ) and Ventilation

The review of design guides and standards on the approaches to cooling and heating, thermal comfort and ventilation in educational facilities indicated that most guidelines, policies and protocols follow the American *ASHRAE Standards 55* and *62.1* and the European *ISO 7730, CIBSE Guide A* and *CR 1752*. The thermal comfort standards (*Standard 55, ISO 7730, CIBSE Guide A*) prescribe numeric and descriptive criteria for comfort primarily for mechanically conditioned buildings—19.7 to 26.7 °C, 20 to 60% RH. The guidance provided for naturally ventilated spaces by *Standard 55* applies only to conditions where the mean monthly outdoor temperature ranges from 10 to 35 °C, and occupants must be able to open and adjust operable windows. Whereas *Standard 55* does not provide specific guidance for naturally conditioned spaces, *CIBSE Guide A* prescribes summer design temperatures and over-heating criteria for free-running buildings, where 25 °C is an acceptable indoor temperature. The criteria and recommendations in the international standards have been adopted as normative references by Australian national standards, guidelines and codes of practice.

Indoor air quality (IAQ) standards pertain to reducing the quantity of indoor air contaminants by providing criteria for ventilation rates. CO_2 concentrations are often used as a surrogate of the rate of outside supply air per occupant, and indoor CO_2 concentrations above about 1000 ppm are generally regarded as indicative of ventilation rates that are unacceptable with respect to body odours [26]. The international standards (*ASHRAE Standard 62, European standards EN 13779* and *CR 1752*) provide both the prescriptive and analytical methods to calculate the ventilation rates. The regulatory actions related to IAQ in Australia are limited, and there is a lack of information on the emissions rates of and exposure levels to pollutants in specific building categories. As an alternative to calculating the concentration levels, exposure to pollutants and actual monitoring, using the ventilation rates, for example those prescribed for educational facilities, is deemed to adequately address the achievement of the required IAQ for a space or building. The ventilation requirements in the *Australian Standard AS 1668.2* for educational facilities (10–12 l/s per person) align with those prescribed in the international standards.

7.2 Policies on Thermal Comfort, Indoor Air Quality (IAQ) and Ventilation for Educational Facilities

In the review of the policies to cooling and ventilation in educational facilities, it is observed that two streams of approaches are typically adopted. General requirements for teaching and learning spaces in North America and the UK are to comply with the recommended performance standards for school buildings where the prescribed indoor temperature or temperature range, ventilation rates and CO_2 concentration levels are met using the standards. In Australia, the states of South Australia and Queensland follow this approach and have outlined the conditions of occupied school spaces which require the provision of air conditioning. The states of Victoria, New South Wales and Western Australia specify the requirement for cooling based on geographic locations and external (climatic) conditions rather than prescribing the indoor conditions (temperature, air quality) of school spaces.

7.3 Indoor Environmental Conditions and Educational Outcomes

Information on indoor environmental conditions in Australian schools is very limited. Few data and scientific studies on measurements of school environments, particularly on thermal conditions and IAQ are available. Moreover, majority of the studies summarised in this review have been conducted in the northern mid-latitudes. This lack of knowledge poses a concern considering that children, unlike adults, are much more vulnerable, are required to perform work that is not optional and would almost always be new to them [26, 47, 73].

The prescribed conditions and temperature limits recommended by the standards were based on studies which did not take peoples work performance into account. Available peer-reviewed literature and studies on the effects of classroom thermal conditions and air quality on student performance are likewise very sparse. However, the findings of the few research studies summarised in this chapter suggest that increased classroom temperatures can have negative effects on the performance of schoolwork by children. These studies indicate that air quality and temperature were improved by increasing ventilation and cooling. However, assumptions that the results of these studies can be generalised to other developed countries where the climate, classroom conditions, level of education and educational approach are similar to those in the northern mid-latitudes will have to be validated by replicating them in temperate, subtropical, tropical and humid climates.

This chapter establishes the need for a study grounded on addressing the absence of clear documentation on the state of indoor environments in educational facilities in Australia backed by measurements and surveys of temperature, comfort conditions, indoor quality and the relationship between these aspects of indoor environments and student performance. The minimisation of temperature extremes within school buildings and IAQ-related impacts may yield significant educational learning outcomes to Australia's P-12 education sector but as yet there is little evidence to back this proposition.

Acknowledgements An earlier and extended version of this chapter was presented in the 2013 report prepared by the authors for the Victorian Department of Education and Early Childhood Development (DEECD), now Victorian Department of Education and Training (DET), who funded the literature review project.

References

1. ABCB (2016) National construction code - building code of Australia, volume 1—class 2 to 9 buildings. Australian Building Codes Board (ABCB), Canberra
2. ABS (2017) 4240.0—Preschool Education, Australia, 2016, Australian Bureau of Statistics (ABS), Canberra
3. ASHRAE (2009) Indoor air quality guide: best practices for design, construction and commissioning. American Society of Heating, Refrigerating and Air-Conditioning Engineers, Inc. (ASHRAE)
4. ASHRAE (2010) ANSI/ASHRAE Standard 55-2010: thermal environmental conditions for human occupancy. American Society of Heating Refrigerating and Air-conditioning Engineers (ASHRAE), Atlanta, GA
5. ASHRAE (2010) ANSI/ASHRAE Standard 62.1-2010: ventilation for acceptable indoor air quality. American Society of Heating Refrigerating and Air-conditioning Engineers (ASHRAE), Atlanta, GA
6. ASHRAE (2011) ASHRAE Handbook—HVAC Applications. American Society of Heating, Refrigerating and Air-Conditioning Engineers, Inc. (ASHRAE), Atlanta, GA
7. ASHRAE (2016) ANSI/ASHRAE Standard 62.1-2016: ventilation for acceptable indoor air quality. American Society of Heating Refrigerating and Air-conditioning Engineers (ASHRAE), Atlanta, GA
8. ASHRAE (2016) ASHRAE Guideline 10-2016: Interactions Affecting the Achievement of Acceptable Indoor Environments. American Society of Heating, Refrigerating and Air-Conditioning Engineers, Inc. (ASHRAE), Atlanta, GA
9. ASHRAE (2017) ANSI/ASHRAE Standard 55-2017: thermal environmental conditions for human occupancy. American Society of Heating Refrigerating and Air-conditioning Engineers (ASHRAE), Atlanta, GA
10. ASHRAE (2017) ASHRAE Handbook—Fundamentals. SI edn, American Society of Heating, Refrigerating and Air-Conditioning Engineers, Inc. (ASHRAE), Atlanta, GA

11. BoM (2013) Annual climate summary 2012. Australian Government Bureau of Meteorology (BoM)
12. BoM (2017) Annual climate statement 2016. Australian Government Bureau of Meteorology (BoM), viewed 20 Jan 2017. http://www.bom.gov.au/climate/current/annual/aus/
13. BoM and CSIRO (2014) State of the climate 2014. Australian Government Bureau of Meteorology (BoM) and Commonwealth Scientific and Industrial Research Organisation (CSIRO
14. CEN (1998) CR 1752-1998: ventilation for buildings—design criteria for the indoor environment. European Committee for Standardization (CEN)
15. CEN (2007) EN 13779-2007: ventilation for non-residential buildings—performance requirements for ventilation and room-conditioning systems. European Committee for Standardization (CEN)
16. Charles D, Chad D (2002) Assessment of link between productivity and indoor air quality. In: DJ Clements-Croome (ed) Creating the Productive Workplace. Taylor & Francis e-Library, London. pp 107–26
17. Cheryan S, Ziegler SA, Plaut VC et al (2014) Designing classrooms to maximize student achievement. Policy Insights Behav Brain Sci 1(1):4–12
18. Chilcott T (2012) Newman Government claims cool schools not a priority as heat rises in classrooms. The Courier-Mail, 4 Dec 2012
19. CIBSE (2006) Guide A: environmental design, 7th edn, The Chartered Institution of Building Services Engineers (CIBSE), London
20. Clausen G, Bekö G, Corsi RL et al (2011) Reflections on the state of research: indoor environmental quality. Indoor Air 21(3):219–230
21. Clements-Croome DJ (2002) Indoor environment and productivity. In: DJ Clements-Croome (ed) Creating the productive workplace. Taylor & Francis e-Library, London. pp 1–17
22. Clements-Croome DJ (ed) (2002) Creating the productive workplace. Taylor & Francis, London
23. Comcare Australia (2008) Occupational health and safety code of practice 2008 Comcare Australia, Canberra, ACT, 2 June 2008. http://www.comlaw.gov.au/Details/F2008L02054
24. Comcare Australia & Community and Public Sector Union (Australia) (1995) Air conditioning and thermal comfort in Australian public service offices: an information booklet for health and safety representatives. Comcare for and on behalf of Australian public service agencies and the Community and Public Sector Union, Canberra, ACT
25. Csobod E, Annesi-Maesano I, Carrer P et al (2014) Final report: Schools Indoor Pollution and Health Observatory Network in Europe (SINPHONIE). European Commission: Directorate General for Health and Consumers and Directorate General Joint Research Centre—Institute for Health and Consumer Protection, Ispra (VA), Italy
26. Daisey JM, Angell WJ, Apte MG (2003) Indoor air quality, ventilation and health symptoms in schools: an analysis of existing information. Indoor Air 13(1):53–64
27. de Dear R (2004) Thermal comfort in practice. Indoor Air 14(Supplement 7):32–39
28. de Dear R, Kim J, Candido C et al (2015) Adaptive thermal comfort in Australian school classrooms. Build Res Inf 43(3):383–398
29. de Dear RJ (1998) A global database of thermal comfort field experiments. ASHRAE Trans 104(1B):1141–1152
30. de Dear RJ, Brager GS, Cooper D (1997) ASHRAE RP-884 final report: developing an adaptive model of thermal comfort and preference. Macquarie Research Limited, Macquarie University and Center for Environmental Design Research, University of California, Sydney, Australia and Berkeley CA, USA
31. DECS (2007) Protocol: SV001: 'air conditioning' (ventilation, heating and cooling) policy. Department of Education and Children's Services (DECS), Asset Policy & Capital Programs, Adelaide. http://www.decd.sa.gov.au/assetservices/pages/topiclisting/hvac/
32. DET and VSBA (2017) Building quality standards handbook. Department of Education and Training (DET) and Victorian School Building Authority (VSBA), Melbourne, VIC
33. DETE (2010) Schools standard air conditioning specification (Version 2. 20 Oct 2010), Queensland Government Department of Public Works for the Department of Education, Training and Employment (DETE), Brisbane. http://deta.qld.gov.au/corporate/design-standards/design-standards-dete-facilities.html

34. DETE (2012) Design standards for DETE facilities: section 3—ecologically sustainable development requirements. Queensland Government Department of Education, Training and Employment (DETE), Brisbane
35. EPA (2009) Indoor air quality tools for schools—reference guide (EPA 402/K-07/008), U.S. Environmental Protection Agency (EPA), Washington, DC
36. Fang L, Wyon DP, Clausen G et al (2004) Impact of indoor air temperature and humidity in an office on perceived air quality, SBS symptoms and performance. Indoor Air 14(Supplement 7):74–81
37. Fanger PO (1972) Thermal comfort: analysis and applications in environmental engineering. McGraw-Hill Book Co., New York
38. Fisk WJ, Satish U, Mendell MJ et al (2013) IAQ applications: is CO_2 indoor pollutant? ASHRAE J 55(3):84–85
39. Francisco P (2011) IAQ applications: weatherization and indoor air quality. ASHRAE J 53(5):96–100
40. Institute for Health and Consumer Protection - European Commission Joint Research Centre (2011) SINPHONIE: turning research into action—recommendations for a healthy school environment in Europe. European Commission Joint Research Centre - Institute for Health & Consumer Protection, Szentendre, Hungary. http://www.sinphonie.eu/sites/default/files/publications/final/Final%20brochure%20for%20schools.pdf
41. ISIAQ-CIB Task Group TG42 (2004) Performance criteria of buildings for health and comfort (CIB number 292, 2004). International Council for Research and Innovation in Building and Construction (CIB) and International Society of Indoor Air Quality and Climate (ISIAQ), Denmark
42. ISO (2005) EN ISO 7730: Ergonomics of the thermal environment—analytical determination and interpretation of thermal comfort using calculation of the PMV and PPD indices and local comfort. International Standards Organisation (ISO), Geneva. http://www.iso.org/iso/catalogue_detail.htm?csnumber=39155
43. Jackson W, Argent R, Bax N et al (2017) Australia state of the environment 2016: overview, independent report to the Australian Government Minister for the Environment and Energy, Australian Government Department of the Environment and Energy, Canberra. Australian Government Department of Environment and Energy, Canberra, ACT
44. Lan L, Wargocki P, Lian Z (2012) Optimal thermal environment improves performance of office work. REHVA J 2012:12–17
45. Luther M, Atkinson SE (2012) Measurement and solutions to thermal comfort, CO_2 and ventilation rates in schools. Paper presented to healthy buildings 2012: Proceedings of the 10th International Conference of Healthy Buildings (8–12 July 2012), Brisbane, QLD
46. Luther M, Horan P (2014) Investigating and understanding CO_2 concentrations in school classrooms. Paper presented to across: Architectural Research through to Practice: 48th International Conference of the Architectural Science Association 2014, University of Genoa, Genoa, Italy (10–14 Dec 2014)
47. Mendell MJ, Heath GA (2005) Do indoor pollutants and thermal conditions in schools influence student performance? A critical review of the literature. Indoor Air 15(1):27–52
48. Mors St, Hensen JLM, Loomans MGLC et al (2011) Adaptive thermal comfort in primary school classrooms: Creating and validating PMV-based comfort charts. Build Environ 46(12):2454–2461
49. Mumovic D, Davies M, Ridley I et al (2009) A methodology for post-occupancy evaluation of ventilation rates in schools. Build Serv Eng Res Technol 3(2):143–152
50. Nevins RG, Gonzalez RR, Nishi Y et al (1975) Effect of changes in ambient temperature and level of humidity on comfort and thermal sensations. ASHRAE Trans 81(2):169–182
51. NSW Department of Education & Communities (2011) Media release: trial for greener, cooler school. New South Wales (NSW) Government Department of Education & Communities, viewed 25 May 2013. http://www.dec.nsw.gov.au/about-us/news-at-det/media-releases1/media-releases-2011/trial-for-greener-cooler-school

52. NSW Department of Education & Communities (2012) Air cooling in schools. New South Wales (NSW) Government Department of Education & Communities, Sydney, NSW
53. NSW Department of Education & Communities (2012) Stronger NSW public schools in 2013: thermal comfort and sustainability standards. New South Wales (NSW) Government Department of Education & Communities, Sydney, NSW. https://www.det.nsw.edu.au/media/downloads/our-services/schools/back-to-school-2013/facts-and-figures/thermal-comfort.pdf
54. Olesen BW (2004) International standards for the indoor environment. Indoor Air 14(Supplement 7):18–26
55. Prescott K (2001) Environment design guide DES 42: thermal comfort in school buildings in the tropics. Royal Australian Institute of Architects
56. Schoer L, Shaffran J (1973) A combined evaluation of three separate research projects on the effects of thermal environment on learning and performance. ASHRAE Trans 79(1A):97–108
57. Simonson CJ, Salonvaara M, Ojanen T (2002) The effect of structures on indoor humidity—-possibility to improve comfort and perceived air quality. Indoor Air 12(4):243–251
58. Spengler JD (2012) Climate change, indoor environments, and health. Indoor Air 22(2):89–95
59. Standards Australia (2012) AS 1668.2-2012: the use of ventilation and airconditioning in buildings—Part 2: mechanical ventilation in buildings. SAI Global under licence from Standard Australia, Sydney
60. Standards Australia (2012) AS 1668.4-2012: the use of ventilation and airconditioning in buildings—Part 4: natural ventilation of buildings. SAI Global Limited under licence from Standards Australia, Sydney
61. Steffen W, Alexander D, Rice M (2017) 2016: global heat record broken again. Climate Council of Australia Sydney, NSW
62. Steffen W, Stock A, Alexander D et al (2017) Angry summer 2016/17: Climate Change Supercharging Extreme Weather Climate Council of Australia Sydney, NSW
63. Sundell J (2004) On the history of indoor air quality and health. Indoor Air 14(Supplement 7):51–58
64. Sundell J, Levin H, Nazaroff WW et al (2011) Ventilation rates and health: multidisciplinary review of the scientific literature. Indoor Air 21(3):191–204
65. Teli D, James PAB, Jentsch MF (2013) Thermal comfort in naturally ventilated primary school classrooms. Build Res Inf 41(3):301–316
66. Teli D, Jentsch MF, James PAB et al (2012) Field study on thermal comfort in a UK primary school. Paper presented to 7th Windsor Conference, Cumberland Lodge, Windsor, UK, 12–15 Apr 2012
67. Toftum J, Fanger PO (1999) Air humidity requirements for human comfort. ASHRAE Trans 105(2):641–647
68. Victoria Department of Education and Training (2017) School policy advisory guide—school hours. viewed 14 Nov 2017. http://www.education.vic.gov.au/school/principals/spag/management/Pages/hours.aspx
69. VTHC OHS Unit (2011) Offices: temperature and humidity—what are the "rules"? Victorian Trades Hall Council (VTHC), viewed 12 July 2013. http://www.ohsrep.org.au/your-industry/education/offices-temperature-and-humidity-what-are-the-rules/index.cfm
70. WA Department of Education (2011) Petition No 68—Air Conditioning in State Government Schools (24 June 2011), Perth, WA
71. WA Department of Education (2013) Schools are cool in 2013! ED-e-News, viewed 18 June 2013. http://det.wa.edu.au/edenews/detcms/corporate-communications-marketing/ed-e-news/news-items/february-2013/schools-are-coolin-2013-.en?oid=NewsItem-id-13670756
72. Wang X, McAllister RR (2011) Adapting to heatwaves and coastal flooding. In: Cleugh H, Smith MS, Battaglia M, Graham P (eds) Climate change: science and solutions for Australia. CSIRO Publishing, Collingwood, VIC, pp 73–84
73. Wargocki P, Wyon DP (2006) Effects of HVAC on student performance. ASHRAE J 48(10):23–27
74. Wargocki P, Wyon DP (2013) Providing better thermal and air quality conditions in school classrooms would be cost-effective. Build Environ 59:581–589

75. Weinstein CS (1979) The physical environment of the school: a review of the research. J Educ Res 49(4):577–610
76. Williamson TJ, Grant E, Hansen A et al (2009) An investigation of potential health benefits from increasing energy efficiency stringency requirements—building code of australia volumes one and two. Adelaide Research and Innovation Pty Ltd., The University of Adelaide, Adelaide SA, Australia
77. WorkSafe Victoria (2006) Officewise—a guide to health & safety in the office (Edition No. 5, Jan 2006), WorkSafe Victoria, Melbourne
78. WorkSafe Victoria (2008) Compliance code: workplace amenities and work environment (Edition No. 01, Sept 2008), WorkSafe Victoria, Melbourne
79. Wyon DP (2004) The effects of indoor air quality on performance and productivity. Indoor Air 14(Supplement 7):92–101
80. Wyon DP, Wargocki P (2009) Indoor environmental effects on the performance of school work by children (RP-1257). International Centre for Indoor Environment and Energy (ICIEE). Department of Mechanical Engineering (MEK) and Technical University of Denmark (DTU), Atlanta, GA
81. Wyon DP, Wargocki P (2013) How indoor environment affects performance. ASHRAE J 55(3):46–53
82. Young E, Green HA, Roehrich-Patrick L et al (2003) Do K-12 school facilities affect education outcomes?. The Tennessee Advisory Commission on Intergovernmental Relations, Nashville, TN

Dr. Mary Myla Andamon, B.Sc. Arch., Ph.D. is a lecturer and a building scientist with expertise in indoor and outdoor thermal comfort studies, urban microclimate and building environmental monitoring and performance assessments.

Dr. Jin Woo, B.S., M.S., M.P.D., Ph.D., Green Star AP is a lecturer in building science with expertise in post-occupancy evaluation of buildings.

University Buildings: The Push and Pull for Sustainability

Matthew Francis and Trivess Moore

Abstract Universities are a key stakeholder in our built environment with buildings in many major cities around Australia and the world. Due to their primarily urban locations, size and number of staff and students, universities and their activities are a significant contributor of greenhouse gas emissions. Increasingly universities both in Australia and globally are looking for ways to improve their sustainability outcomes. This recognizes that higher education institutions can do more to help in the transition to a low-carbon future, but also that by adopting sustainability initiatives, universities help reduce operating costs and facilitate healthier and more productive staff and students. This chapter explores the role of universities and their sustainability initiatives including their challenges of servicing complex stakeholders in a transition to a low-carbon future. After discussing relevant policies and rating tools, five key examples that go significantly beyond minimum performance requirements from prominent Australian universities are presented. Evident from the examples is that there continues to be no one-size-fits-all approach for universities to become more sustainable. It will require complex considerations of the requirements of the university anticipated future needs as well as a wide-ranging evaluation of the most appropriate pathways forward. Ultimately, it is encouraging to see key universities engaging more seriously with improving sustainability outcomes, not only in Australia but also globally. Universities have the opportunity to not just improve sustainability of their facilities, but to also demonstrate to their hundreds of thousands of students and staff how the built environment can be designed to benefit both the environment and the occupants.

M. Francis (✉) · T. Moore
Sustainable Building Innovation Laboratory, School of Property Construction and Project Management, RMIT University, Melbourne, VIC, Australia
e-mail: matthew.francis@rmit.edu.au

T. Moore
e-mail: trivess.moore@rmit.edu.au

P. Rajagopalan et al. (eds.), *Energy Performance in the Australian Built Environment*, Green Energy and Technology, https://doi.org/10.1007/978-981-10-7880-4_9

1 Introduction

Globally, universities are important organizations within our societies for teaching the next generation and delivering critical research to help address societal problems (e.g. see Chapters "The Built Environment in Australia" and "The Built Environment and Energy Efficiency in Australia: Current State of Play and Where to Next"). Most universities in Australia have their physical presence in cities. Owing to increasing growth in the sector, this presence is rapidly expanding through an increase in the occupation and operation of their physical capital, that is, buildings used for teaching, research and administrative purposes.

Universities due to growth in the higher education sector both in Australia and globally have identified an increased pressure on their campus facilities to facilitate the increasing teaching and research expectations of students, government, academia, industry and the tax-paying public. Within the Australian context, this pressure is attributed to increasing trends towards greater financial accountability and independence [16, 20, 21], greater neo-liberalization of higher education [50], and public sector management accountability [5, 10, 36]. Student numbers are also increasing with staff–student ratios increasing from 1:12 in 1990 to approximately 1:21 in 2013 [45]. Despite universities being increasingly covered with layers of virtual technology to disseminate learning beyond the physical campus, the physical campus plays an important role in ongoing intellectual engagement and community-building of future ideas and relationships.

A requirement for improving learning outcomes and increasing staff and student well-being are inter-related with the performance of the buildings and facilities that house these processes [3, 7, 12, 14, 27]. The needs, purpose and options for creating these facilities to enable improved sustainable environmental futures will be discussed throughout this chapter. This chapter seeks to contribute further knowledge on university campus sustainability initiatives building on research developments in the past two decades [4, 9, 28, 30, 31, 33, 34, 38]. It does so by presenting a snapshot of Australian universities and their associated built environment characterizations, alongside some notable innovative case studies. The chapter concludes with remarks and recommendations for future research opportunities.

2 Universities in Australia

Universities are found in every major Australian city and most major regional centres with 43 registered universities as of 2017 [8]. The broader education sector is currently considered a central pillar of Australia's modern economy and is the country's third largest export, bringing in around AUD$21.8 billion in 2016 (Fig. 1).

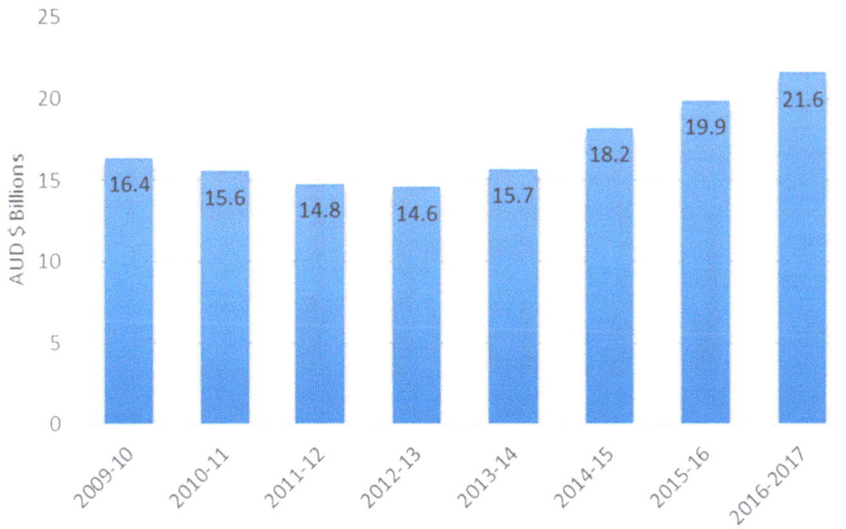

Fig. 1 Australian export income from education services [23, 24]

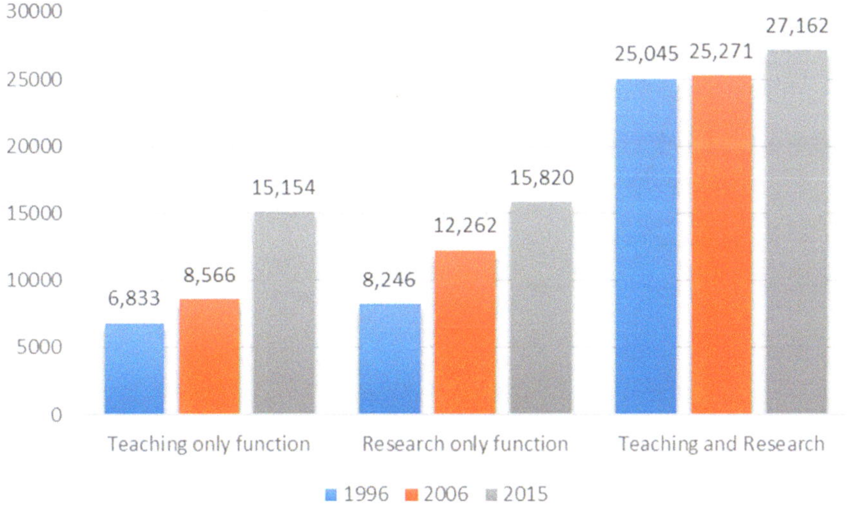

Fig. 2 Number of total Australian university full-time equivalent casual and permanent staff by function 1996, 2006, 2015 [45]

Furthermore, the education sector employs approximately 8% of Australian workers [1, 13]. Universities, as a key part of this education sector, are employing an increasing number of both academic and professional support staff (Fig. 2) with current figures equating to greater than 55,000 full-time equivalent academics, with a combined total of over 160,000 full-time equivalent professionals [19].

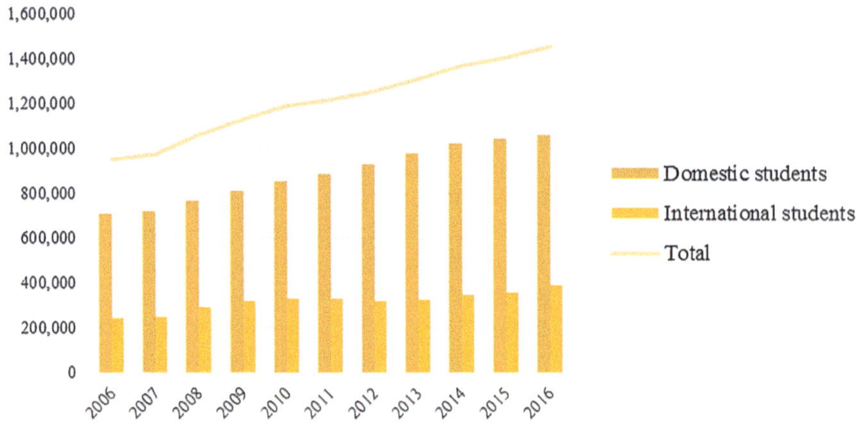

Fig. 3 Total annual student enrolments in Australian higher education institutions, 2006–2016 [23, 24]

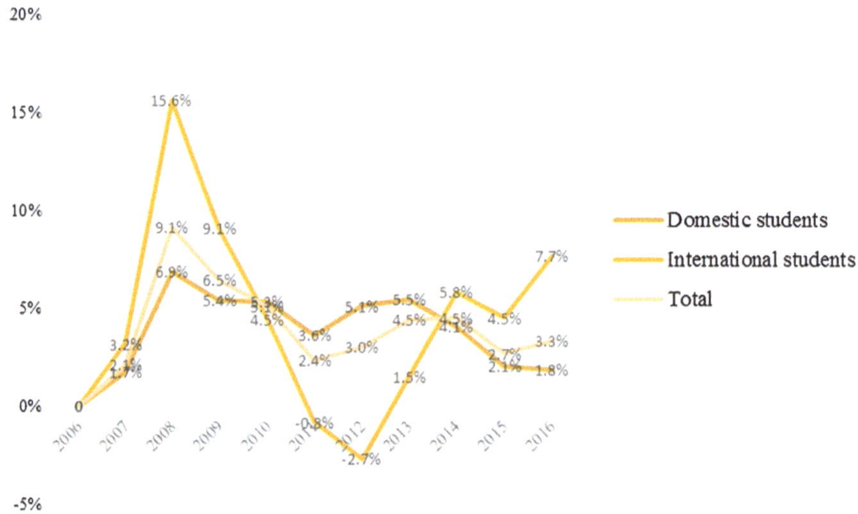

Fig. 4 Growth rates of total student enrolments in Australian higher education institutions from each subsequent year, 2006–2016 [23, 24]

By the end of 2016, there were over 1.45 million domestic and international students enrolled in higher education institutions (Fig. 3). With only slight reductions in overall numbers during 2011 and 2012 in international student enrolments, the average annual growth over the past 10 years was 4.2% for domestic students and 4.8% for international students (Fig. 4).

Accommodating this increasing number of staff and students means a significant amount of financial investment for new, and for the retrofitting of existing, university

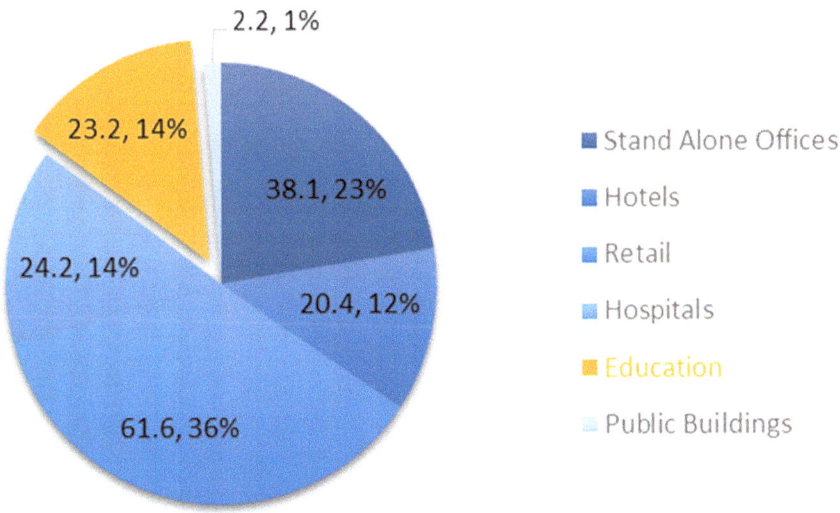

Fig. 5 Total Australian non-residential energy consumption by building type in 2020 (PJ, % of total) [17]

buildings. The quantity of current investment in property development and operation equates to over AUD$3 billion annually within the university sector [20–22]. With such a significant built environment investment year on year and with strong upward trends towards greater expansion, there is opportunity for universities to deliver more sustainable, energy-efficient built environments for academic research staff and students.

With current trends in development and growth in Australia, universities are unsurprisingly a significant contributor to greenhouse gas emissions, primarily through increasing fossil fuel energy consumption. In 2014, the tertiary sector was estimated to account for 8% of Australia's total greenhouse gas emissions [20, 21]. University buildings in Australia are responsible for 10.5 PJ of energy use per annum (as of 2017), equivalent to 6.6% of total non-residential building energy [39]. It has been estimated that energy consumption in the tertiary education sector within Australia will rise to 20.4 PJ in 2020, representing around 14% of total non-residential energy consumption from the built environment (Fig. 5)[1] [17].

This increase in energy consumption coincides with the increasing rise in gross floor square metre area (GFA) of tertiary building stock across Australia, which is expected to rise to 12,000,000 m^2 by 2020 (Fig. 6).

Within universities, often the most energy-intensive spaces are laboratories, cafes and lecture theatres [17]. There is also emerging evidence to indicate that university office buildings lack 'utilization' and, despite being continually mechanically conditioned, often remain unoccupied for periods of time [15]. Although consistent with

[1]This percentage is for the Education sector as a total including for tertiary education buildings.

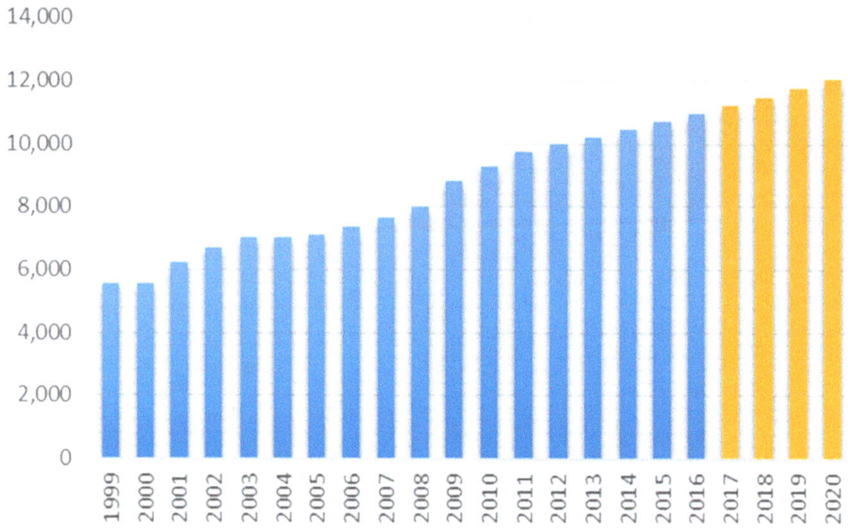

Fig. 6 Total Australian tertiary building area, GFA 1999–2020 [17]

aforementioned trends towards greater cost accountability and neo-liberalization of the university sector, the need for academic staff to contribute to teaching and learning outcomes often requires working in other locations outside of their offices. Therefore, utilization is a flawed metric for building efficiency from an academic office provision perspective.

The majority of emissions can be attributed to the provision of indoor environment quality (IEQ) in spaces used for staff and students such as lighting, and space heating and cooling [42]. Improving systems that are linked to providing IEQ for new and existing university buildings is an important method of reducing the overall environmental impact and ecological footprint of these campuses [40].

Despite little consistency in approaches taken by higher education institutions to environmental target setting, implementing strategies and reporting on progress of sustainability initiatives [9], opportunities exist for universities in Australia to improve their sustainability performance, and in particular their overall energy performance. By seeking to balance their space provisions and utilization commitments while in parallel targeting the provision of environmental conditions within their various built environment configurations, such reductions would play an important role in achieving Australia's carbon reduction goals and renewable energy targets [32].

3 Moving to Sustainable Universities in Australia

There is increasing engagement by universities around the world to improve their sustainability performance by reducing greenhouse gas emissions and improve energy, waste and water efficiency. For example, following the UN Framework Convention on Climate Change in Paris in 2015 [48], more than 300 American universities and colleges pledged to 'accelerate the transition to low-carbon energy while enhancing sustainable and resilient practices across our campus(es)' [43]. This commitment to sustainability and the American Campuses Act on Climate Pledge was reaffirmed by a number of prominent universities [29] following the newly elected US government changing their intent regarding the Paris Agreement [44].

In Australia, the majority of Australian universities have now committed to reducing their greenhouse gas emissions in line with Australia's commitment to reduce its impact by 5% by 2020 [2]. However, a number of universities have set targets for improvement beyond this minimum. For example, a group of five prominent universities (known as the Australian Technology Network) announced a joint commitment in 2009 to collectively reduce greenhouse gas emissions from a base year of 2007 by 25% by 2020, reducing greenhouse gas emissions intensity to 105 kg CO_2^{-e}/m^2/year [41].

Improvement in the energy efficiency of university buildings in Australia has also been assisted through adoption and disclosure of several key voluntary and mandatory tools and regulations including NABERS (National Australian Built Environment Rating System) and widespread adoption of Green Star ratings (as discussed in Chapters "The Built Environment and Energy Efficiency in Australia: Current State of Play and Where to Next" and "Environmental Rating Systems for Non-residential Buildings—How Does Australia Compare with International Best Practice?"). Heightened minimum building performance regulations (including energy requirements) implemented in 2006 and 2010 have also likely improved the implementation of energy efficiency in university buildings. Other countries have similar tools which are applicable to a range of buildings including university buildings, with more notable examples including LEED (USA) and BREEAM (UK).

In relation to the Green Star rating tool, the Green Building Council of Australia has been supportive of universities seeking to adopt their sustainability rating scheme for new and existing buildings with now over 90 university projects now having achieved a formal Green Star rating (Table 1).

While universities are increasingly engaging both formally and informally with the Green star rating system to help improve sustainability outcomes of their buildings, the National Australian Built Environment Rating System (NABERS), has not seen significant uptake by universities. NABERS is a federal government initiative to measure and compare the environmental performance of Australian buildings. It is primarily focused on establishing energy, water, waste and indoor environment quality benchmark measures for buildings specifically created entirely of office environments. Although offices predominate, there have been recent additions to include business hotels, data centres and landlord's services in shopping centres. NABERS

Table 1 Green Star certification data of university projects by state (GBCA 2017)

Rating tool	ACT	NSW	QLD	SA	TAS	VIC	WA	Total
Green Star—Communities						1	1	2
Green Star—Education (2007–2014)			1			2		3
Green Star—Education As Built (2009–2014)		2	5	3		4		14
Green Star—Education Design (2009–2014)	1	14	6	4	2	23	1	51
Green Star—Interiors (2007–2017)							1	1
Green Star—Multi Unit Residential As Built						2		2
Green Star—Multi Unit Residential Design		1			1	4		6
Green Star—Office As Built	1	2						3
Green Star—Office Design	2	4				2		8
Green Star—Public Building As Built						1		1
Green Star—Public Building Design						1		1
Grand Total	4	23	12	7	3	40	3	92

may be used to measure energy efficiency at three levels: base building, tenancy and whole building; however, it is primarily used to measure the base building energy consumption that does not include the energy used by the tenants or take into account occupant densities (for buildings comprising of office areas greater than 75% of total net lettable area). Increasingly, government departments have stipulated minimum ratings for offices they occupy. Following the Building Energy Efficiency Disclosure Act 2010 Commercial Building Disclosure (CBD) programme, the federal government has required that all sellers or lessors of office space greater than 1000 m^2 net lettable area to obtain and disclose the subsequent Building Energy Efficiency Certificate (BEEC) when advertising or undertaking a transaction of leasing or selling the space [25].

Should a building within the university's portfolio seek a NABERS rating assessment, the building must effectively be used for only office workplace provisions and associated facilities. If a university building be a combination of office, lecture theatre, laboratory, etc., the NABERS rating is not wholly appropriate without significant calculation to determine the exact percentage of the total building is attributable to office-related environments. As of August 2017, the annually updated NABERS register of buildings Australia-wide that have sought a formal rating that was identi-

Table 2 University office buildings rated under NABERS (as of August 2017) (Office of Environment & Heritage 2017)

University	Address	NABERS Energy star rating
University of Wollongong	Wollongong, NSW	5.0 stars (base building)
Victorian TAFE Association	Melbourne, VIC	5.0 stars (tenancy)
Melbourne Institute of Technology	Sydney, NSW	4.5 stars (base building)
Australian Catholic University	Sydney, NSW	3.5 stars (base building)
University of New South Wales	Canberra, ACT	2.0 stars (base building)

fied as a university (or higher education) building was limited to only 5 as shown in Table 2. This indicates that there is opportunity for government to encourage university building developments to adopt a NABERS rating as part of future sustainability initiatives.

4 Case Studies and Discussion of Sustainability Innovation in Australian Universities

While the aforementioned approaches and tools are helping to deliver more sustainable universities, there is a need to do more in the immediate future if we are to meet broader societal sustainability goals, such as the UN sustainable development goals [47]. In addition, universities are finding themselves in a rapidly changing environment both in terms of teaching delivery (e.g. moving to more online delivery) and sourcing of funding for facility development. It will take additional innovation beyond the above existing measures if sustainability targets are to be met, if improvements for staff and student conditions and productivity are to be achieved and if it is to be undertaken within heightening financial constraints. There are a number of innovation examples from Australian universities which may inspire other universities, both in Australia and globally. Below we briefly discuss five recent examples.

4.1 Case Study 1—Charles Sturt University—Carbon Neutral Offsets

Charles Sturt University has campuses across three states in Australia (New South Wales, Queensland and Victoria) with more than 40,000 students in both rural and urban locations. The university was the first Australian university to be certified as a carbon neutral university in July 2016 against the National Carbon Offset Standard as well as the first to adopt in parallel concepts of indigenous ideas of sustainability. This carbon neutrality standard was not achieved quickly, with the target first included in the future strategy of the university in 2007. This was as part of demonstrating that the university exists 'for the public good' and, further, sought to promote and adopt local indigenous awareness of sustainability concepts such as 'Yindyamarra Winhanganha' ('the wisdom of respectfully knowing how to live well in a world worth living in') [18].

In the base year of 2014, the total emissions for the university were approximately 43,500 tonnes of CO_2^{-e} [18]. The university having decided to reduce their environmental impact sought energy efficiency improvements using local and international offsets. This allowed for local benefits such as restoration of local fauna and supporting local businesses, while the international benefits included helping developing countries through the provision of more reliable and sustainable energy.

The university is reporting savings of $500,000 per year as a direct result from energy efficiency and renewable energy generation measures [18]. They are also reporting that there are broader benefits for the university such as helping retain and attract staff, improving the comfort and amenity of facilities and better learning and teaching outcomes from improved sustainability. Key learnings as reported by the university includes needing to improve monitoring and measurement of current performance before working out mechanisms to improve sustainability, and that even small changes will add up to make a big difference [18].

4.2 Case Study 2—Monash University—Net Zero Carbon Project

Monash University has a student population of more than 73,000 students which are located over 7 campuses with five in Victoria and two international (Malaysia and South Africa). The university has recently announced that it will spend AUD$135 million to 2030 to transition to 100% renewable energy and target Passive House certification for university buildings [37]. This builds upon their previous sustainability energy target of a 20% energy reduction and the development of 1 MW of solar energy. The university currently has a requirement for all new buildings over $100 million to achieve a sustainability rating.

The new sustainability plan will involve removing gas completely from their fuel mix, the development of a microgrid, use of renewable energy (an additional 3 MW of solar are planned), and storage and energy reduction approaches. Onsite renewables are expected to cover up to 20% of total energy requirements with the remaining renewable energy to be procured through a power purchase agreement with a large-scale renewable energy supplier. The target will not just be bound to the physical campus but include emissions from all operations including plane travel. The university wants to show that improved sustainability is good for the environment and for business.

4.3 *Case Study 3—RMIT University—Energy Performance Contracting*

RMIT University has a student population of more than 78,000 across several local and international campuses (e.g. Vietnam and Spain). In 2010, RMIT University committed AUS$128 million to cut energy, water use and greenhouse gas emissions. Working across three campus locations over two years, opportunities for energy and water savings in 77 buildings (of more than 120 buildings total) were identified and included into an Energy Performance Contract (EPC) with two major international EPC companies. An EPC is essentially a contract for a set improvement in performance to be achieved, or the company with whom the contract is with pays a penalty for failing to achieve the established targets.

Under the title of the 'Sustainable Urban Precincts Program' (SUPP), a reduction in electricity use at RMIT University was estimated to be 263 million kWh (over 8 years), leading to an approximate 32,000-tonne reduction in greenhouse gas emissions (from approximately 78,000 tonnes in 2015). Water use reductions were estimated at 53 million L/year. The energy-saving initiatives focused on: energy-efficient lighting (upgrades to LED), solar renewable energy generation, consolidated building management systems, water saving measures and energy-efficient heating and cooling assets.

An integrated approach to investment in capital and operational requirements provided improved mechanical infrastructure conditions, increased operational efficiencies and tangible sustainability outcomes. The commencement of the first Energy Performance Contract for one campus was in July 2017 with the remaining two campuses' contracts starting January 2018 (Fig. 7).

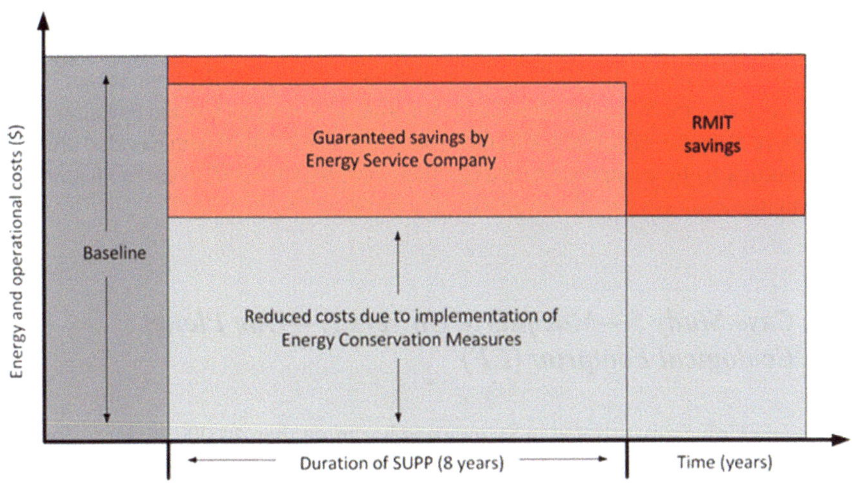

Fig. 7 Energy Performance Contract conceptual diagram (*source* [6])

This above approach has not typically been applied in the university sector in Australia but has the ability to help improve sustainability outcomes while controlling performance outcomes and financial risk. This will be important for universities moving forward, especially as more stringent sustainability targets are introduced, by not only universities, but also governments.

4.4 Case Study 4—University of New South Wales—Live Consumption Feedback

The University of New South Wales (UNSW) has a main campus in Sydney, with two other local campuses [49]. The university has a student population of more than 50,000 students. UNSW is implementing a range of sustainability approaches, including a focus on improving data collection and provision of live feedback and information for building users. This was in recognition that information and benchmarking progress is important if everyone at the university was to engage in a meaningful way with a reduction in resource consumption. The university points to research to support their approach which shows energy savings of 5–15% when users are provided with direct and live information about consumption [26].

In 2012, the Energy Management Unit at UNSW partnered with Greensense, a company that extracts the sustainability performance of a building and presents it in real time via easy-to-understand formats. The system uses more than 650 smart meters to track energy, water and gas consumption across the university. This is useful for not only evaluating how buildings are performing or being used, but also as a way of identifying faults or wastage.

While some of this data was collected previously, it was only accessible by a small number of university staff. Now the data is presented on dashboards online and in each building. By opening the data to be accessible not only by any staff or student (and the wider public), but providing information in real time the hope is that it will make occupants more accountable for their resource consumption and consider ways to improve resource efficiency. The more detailed collection of data has also allowed for buildings to be compared or shown their progress towards a more sustainable performance.

4.5 Case Study 5—Macquarie University—One Planet Ecological Footprint (EF)

Macquarie University is based in Sydney and has more than 40,000 students. Recognizing that if student and staff growth expectations were to be achieved and that this would increase their environmental impact, Macquarie University has been implementing a range of sustainability goals and approaches over the past decade. This has come with the desire to achieve environmental energy-use performance within the resource limitations and bounds of one planet. The One Planet Living ecological footprint (EF) calculation is developed by the Bioregional organization founded in 1994 and known for their BedZED eco-village [11]. The university's goal is to achieve a one planet performance (or better) by 2030.

Working with the Footprint Company, the university first developed a benchmark of performance to understand how they compared to the one planet goal. The university's main campus footprint has tracked between 1.3 and 1.4 planets. To achieve the one planet goal will require at least a 25% improvement in performance by existing buildings. To achieve that the university has a number of specific strategies including to meet 5-star Green Star performance and 4.5-star NABERS energy and water rating for all buildings [35].

5 Implications for Policy and Practice

Universities are inherently places of learning that seek to beneficially impact individuals, society and, increasingly, the environment. It is contended that universities are in a unique position to address ways of learning to live in a manner that does not endanger the earth [46]. Individually and collectively, universities may make significant progress towards leading new generations of students in research endeavours seeking paradigm shifts in understanding energy efficiencies and sustainable energy production, as well as becoming more sustainable in their own operations.

As discussed in this chapter, current benchmarking paradigms for sustainability are limited in their usefulness in order to achieve the levels of energy and overall CO_2 reductions that are necessary to fulfil current sustainability targets. However, there are increasing examples of Australian universities who are taking this challenge on directly with varying levels of success, with or without building rating systems.

What is clearly evidenced in the aforementioned case studies, as well as other notable examples around the world, is that universities are unique in terms of their locations, buildings, uses, students, staff, growth rates, etc., and so there is no 'cookie-cutter' response which will be suitable for the majority of universities. There are of course common areas of sustainability to target which include operational energy and water in both teaching and office spaces. Some of these 'easy' efficiency improvements also help universities to reduce their operating costs for limited capital costs.

However, once these easier sustainability outcomes are achieved, universities are making different decisions based upon a range of complexities such as existing buildings, future projects of students and staff as well as needing to consider constantly changing roles of universities (e.g. shifting to more online teaching delivery). The examples explored in this chapter show there are a range of options a university could take. For example, Charles Sturt University has implemented a combination of purchasing offsets and increasing renewable energy generation as a key strategy, whereas Monash University is going to include a significant focus on improving the performance of new and existing buildings as well as rapidly increasing renewable energy generation and moving to its own microgrid. However, they are also trying to go beyond just direct environmental impacts at specific campus locations to include elements such as travel.

Each approach has its own risks in terms of costs and what the outcomes will be. Getting the assumptions right in the current and future requirements for universities is a critical element in the planning for more sustainable universities. RMIT University has taken further steps to reduce such risks but putting the financial risk onto a third party to deliver promised efficiency gains. There may be other opportunities for innovations like this to help overcome challenges with capital costs to improve sustainability outcomes.

While universities educate the citizenry with interdisciplinary knowledge, they are also often large, prestigious and influential institutions in their own right, capable of having large impacts on the environment, as well as influence local and global communities. As discussed earlier in this chapter, universities are uniquely positioned to both initiate greater learning on the imperatives associated with sustainable development while in parallel making a significant contribution to the adoption of innovative energy policy and building development. What is required is a global push by universities everywhere to share their experiences and support each other towards a more sustainable future. Universities must not underestimate their ability to help shape a more sustainable built environment.

6 Conclusion

Universities are a key stakeholder in our built environment. Many universities are urban in nature and often have a significant footprint in key locations in our cities. Due to the increasing number of staff and students at many universities, and the significant number of buildings they occupy, they are a significant contributor of greenhouse gas emissions and consumer of resources. Increasingly, universities in Australia and globally are looking for ways to improve their sustainability outcomes. This is both in recognition that higher education institutions can do more to help in the transition to a low-carbon future, and also that by adopting sustainability initiatives, universities may help reduce operating costs and facilitate healthier and more productive staff and students.

In Australia, there are a number of notable examples of universities who are implementing sustainability innovations. Case study examples including Charles Sturt University, Monash University, RMIT University, Macquarie University and the University of Melbourne are presented. What is clear is that, consistent with markets outside of academia, there continues to be no one-size-fits-all approach for universities to become more sustainable. It will require complex considerations of the requirements of the university anticipated future needs as well as a wide-ranging evaluation of the most appropriate pathways forward.

Common solutions that are evident in universities seeking to reduce their overall carbon emissions are the consolidating all offsets and managing these centrally so as to benefit from a centralized pool, maximizing energy efficiency savings through appropriate building and equipment selections, and better energy, water and land management strategies. Building rating systems have been used effectively to help drive sustainability innovations in university campuses, both in terms of policy and building design.

Ultimately, it is encouraging to see key universities engaging more seriously with improving sustainability outcomes, not only in Australia but globally. Universities have the opportunity to not just improve sustainability of their facilities, but to also demonstrate to the hundreds of thousands of students who use these facilities each year what a more sustainable future looks like.

References

1. ABS (2015) Labour force, Australia (6291.0.55.003). Australian Bureau of Statistics, Canberra
2. ACTS (2013) ACTS history. Retrieved May, 2015 from http://www.acts.asn.au/resources/
3. Allen JG et al (2015) Green buildings and health. Curr Environ Health Rep 2(3):250–258
4. Amaral LP et al (2015) Quest for a sustainable university: a review. Int J Sustain High Educ 16(2)
5. Anderson G (2006) Carving out time and space in the managerial university. J Organ Change Manage 19(5):578–592
6. Anderson M (2016) RMIT property services-sustainable urban precincts programme energy performance contracting
7. Arnold H (2014) Why green buildings increase productivity. Retrieved July, 2015 from http://www.afr.com/business/construction/why-green-buildings-increase-productivity-20140716-ixaop
8. Australian Trade and Investment Commission (2017) Universities and Higher Education. Retrieved November, 2017 from https://www.studyinaustralia.gov.au/english/australian-education/universities-and-higher-education
9. Bekmann H et al (2013) Ecological footprint benchmarking of 40 tertiary education campuses. In: Proceedings of the 13th international Australasian campuses towards sustainability (ACTS) conference, Sydney, Australia. http://www.acts.asn.au/index.php/2013-conf/conference-proceedings
10. Bellamy S et al (2003) Why business academics remain in Australian universities despite deteriorating working conditions and reduced job satisfaction: an intellectual puzzle. J High Educ Policy Manage 25(1):13–28
11. Bioregional (2017) One planet living. Retrieved November, 2017 from http://www.bioregional.com/oneplanetliving/
12. Brown Z et al (2010) Evaluating user experience in green buildings in relation to workplace culture and context. Facilities 28(3/4):225–238
13. Coelli M (2015) Australia's 'five pillar economy': education. Retrieved July, 2016 from https://theconversation.com/australias-five-pillar-economy-education-40831
14. Cole R (2003) Green buildings–reconciling technological change and occupant expectations. In: Lorch R, Cole R (eds) Buildings, Culture and Environment: Informing Local and Global Practices. pp 57–82
15. Dane J (2017) New frontiers in designing the academic workplace. Retrieved August, 2017 from https://www.woodsbagot.com/ideas/new-frontiers-in-designing-the-academic-workplace
16. Dawkins J (1987) Higher education: a policy discussion paper. Australian Government Publishing Service

17. DCCEE (2012) Baseline Energy Consumption and Greenhouse Gas Emissions in Commerical Buildings in Australia. Department of Climate Change and Energy Efficiency, Australian Government, Canberra
18. DEE (2017) Carbon neutral stories—Charles Sturt University
19. Deloitte (2015) The importance of universities to Australia's prosperity
20. Department of Education and Training (2015) Finance 2014: financial reports of higher education providers. Department of Education and Training, Canberra
21. Department of Education and Training (2015) Higher education in Australia: a review of reviews from Dawkins to today. Department of Education and Training, Australian Government, Canberra
22. Department of Education and Training (2016) Selected higher education statistics—2015 staff data. Retrieved July, 2016 from https://www.education.gov.au/selected-higher-education-statistics-2015-staff-data
23. Department of Education and Training (2017) Export income to Australia from international education activity in 2016. Australian Government, Canberra
24. Department of Education and Training (2017) Higher education statistics: student data. Retrieved December, 2017 from https://www.education.gov.au/student-data
25. Department of Environment and Energy (2017) Commerical building disclosure: a national energy efficiency program. Retrieved December, 2017 from http://cbd.gov.au/
26. Fischer C (2008) Feedback on household electricity consumption: a tool for saving energy? Energ Effi 1(1):79–104
27. Heerwagen J (2010) Green buildings, organizational success and occupant productivity. Build Res Inf 28(5):353–367
28. Hodge T (1997) Toward a conceptual framework for assessing progress toward sustainability. Soc Indic Res 40(1–2):5–98
29. Hub Report (2017) Johns Hopkins joins 11 other universities in reaffirming commitment to tenets of Paris accord. Retrieved August, 2017 from https://hub.jhu.edu/2017/06/05/hopkins-climate-pledge-joint-statement/
30. Intrachooto S, Arons D (2002) Nurturing green innovations for academic institutions. Int J Sustain High Educ 3(2):155–163
31. Iyer-Raniga U et al (2015) Beyond buildings: holistic sustainable outcomes for university buildings. In: 15th International Australasian Campuses Towards Sustainability, 21–23 October
32. Iyer-Raniga U et al (2016) Aligning goals for sustainable outcomes: case study of a university building in Australia. In: Leal W, Brandli L (eds) Engaging stakeholders in education for sustainable development at university level. London, Springer
33. Leal Filho W et al (2015) The future we want: key issues on sustainable development in higher education after Rio and the UN decade of education for sustainable development. Int J Sustain High Educ 16(1)
34. Luo H, Yang J (2012) Overcoming organisational resistance to sustainability innovations in Australian universities. In: Proceedings of the 12th Annual Australasian Campuses Towards Sustainability
35. Macquarie University (2013) Sustaining our future: Sustainability strategy 2015–2019. Retrieved November, 2017 from https://www.mq.edu.au/about/about-the-university/strategy-and-initiatives/strategic-initiatives/sustainability/media/annual-reports-and-publications/Sustaining-Our-Future_LowRes.pdf
36. Martin-Sardesai A et al (2017) Organizational change in an Australian university: responses to a research assessment exercise. Br Acc Rev
37. Monash University (2017) Net zero initiative
38. Pineno CJ (2011) Sustainability reporting by universities: a separate category within the balanced scorecard based on key drivers through a mapping strategy. J Strateg Innov Sustain 7(2):122–134
39. Pitt & Sherry (2012) Baseline energy consumption and greenhouse gas emissions in commercial buildings in Australia—Part 1—Report. Department of Climate Change and Energy Efficiency, Canberra

40. Rickards S et al (2015) Ecological footprinting as a top-down and bottom-up approach to complete campus engagement and transformation towards the One Planet Goal. Integrative Approaches to Sustainable Development at University Level, Springer, pp 43–66
41. Riedy C, Daly J (2010) Targeting a low-carbon university: a greenhouse gas reduction target for the Australian technology network of universities. In: Filho W (ed) Universities and climate change: introducing climate change to university programmes. Springer, Berlin
42. TEFMA (2015) TEFMA benchmark 2014. T. E. F. M. Association, Tertiary Education Facility Management Association (Australia)
43. U.S. Office of the Press Secretary (2015) Fact sheet: ahead of the conference on climate change, more than 200 colleges and universities sign the American Campus Act on climate pledge to demonstrate support for strong international climate action. Retrieved August, 2017 from https://obamawhitehouse.archives.gov/the-press-office/2015/11/19/fact-sheet-ahead-conference-climate-change-more-200-colleges-and
44. U.S. Office of the Press Secretary (2017) Statement by President Trump on the Paris Climate Accord. Retrieved from https://www.whitehouse.gov/the-press-office/2017/06/01/statement-president-trump-paris-climate-accord
45. UA (2015) Higher education and research: facts and figures 2015. Universities Australia
46. Uhl C, Anderson A (2001) Green destiny: universities leading the way to a sustainable future. AIBS Bull 51(1):36–42
47. UNEP (2015) Sustainability development goals. Retrieved December, 2017 from http://www.un.org/sustainabledevelopment/sustainable-development-goals/
48. United Nations (2015) Framework convention on climate change 'Paris Agreement'. UNFCCC
49. University of New South Wales (2017) Campus development. Retrieved November, 2017 from http://www.estate.unsw.edu.au/live-energy
50. Ward SC (2012) Neoliberalism and the global restructuring of knowledge and education. Routledge

Matthew Francis, B.E., M.E. is a research fellow with expertise in building energy efficiency, particularly in the relationship between building performance and occupant productivity while enhancing workplace sustainability.

Dr. Trivess Moore, B.Sc. (Env.), Ph.D. is a research fellow with expertise in energy efficiency, renewable energies, the built environment and transitioning to a low-carbon future.

A Guide for Evaluating the Performance of Indoor Aquatic Centres

Priyadarsini Rajagopalan

Abstract Aquatic centres are popular recreational facilities in Australia. These buildings have experienced increasing demand over the past few decades. Aquatic centres are complex building types accommodating diverse facilities that are distinct in their functional requirements. The high-energy intensity and growing desire for better indoor environmental quality in aquatic centres have resulted in a marked increase in energy consumption in this sector which presents a great challenge in terms of new construction and renovation. This chapter provides an overview of the characteristics of aquatic centres, highlighting the challenges in evaluating the performance of these buildings. A methodology for evaluating the design and operational performance of these buildings is also proposed.

1 Introduction

The demand for sports facilities in Australia has been increasing in urban areas over recent decades. As a result of this and a general focus on improving sustainability in the built environment (see Chapters "The Built Environment and Energy Efficiency in Australia: Current State of Play and Where to Next" and "Environmental Rating Systems for Non-Residential Buildings—How Does Australia Compare with International Best Practice?"), more attention is drawn towards creating a healthy indoor environment for users. Aquatic centres represent popular recreational and sports facilities in Australia. Aquatic centres are complex building types accommodating diverse facilities such as swimming pools, gymnasiums, fitness centres, sports halls, cafés, crèches and offices that are distinct in their functional requirements. An increase in the number and the use of aquatic centres has seen such multi-purpose indoor recreational facilities become vital points for community interaction. Aquatic

P. Rajagopalan (✉)
Sustainable Building Innovation Laboratory, School of Property Construction and Project Management, RMIT University, Melbourne, VIC, Australia
e-mail: priyadarsini.rajagopalan@rmit.edu.au

P. Rajagopalan et al. (eds.), *Energy Performance in the Australian Built Environment*, Green Energy and Technology, https://doi.org/10.1007/978-981-10-7880-4_10

centres have high-energy intensity, and this presents a great challenge in terms of new construction and renovation.

In Europe, sports buildings with swimming pools are seen as potential targets for reducing building energy use because of their high energy consumption. This is in light of the '2030 Climate and Energy Framework', aimed at a reduction target of the greenhouse gas emissions of at least 40% by 2030 compared to 1990 [11]. Every year, £700 million is spent on energy by sports sector buildings, and this is equivalent to annual emissions of 10 million tonnes of carbon dioxide [8]. According to Carbon Trust [9], energy costs account for nearly 30% of the total running costs of any typical sports centre and are second only to labour costs.

The aquatic and recreation industry in Australia has over 250 million visitors per year and employs over 86,000 staff [5]. According to Aquatics and Recreation Victoria (ARV), there are around 1900 aquatic centres in Australia of which 500 aquatic facilities are located in Victoria. Among the Victorian centres, 277 belong to the local government. The remaining 233 are owned by private swim schools or educational institutions. During the early 1970s, Australian state and local governments implemented many programmes introducing new sports and aquatic facilities with the provision of indoor activities. Due to the cold weather conditions in Victoria, water bodies were needed to be enclosed and as a result Victoria has the largest number of indoor centres in Australia. Other Australian states with more subtropical/tropical climates have traditionally used open-air infrastructures or interactive spray parks and play spaces to benefit from the warmer climate.

This chapter provides an overview of the characteristics of aquatic centres, highlighting the challenges in evaluating the performance of these buildings. The findings are informed by detailed study of selected aquatic centres in Victoria. The problems arising from the lack of clear definitions are discussed, and a general methodology for evaluating the design and operational performance for aquatic centres is provided.

2 Challenges in Benchmarking Aquatic Centres

Aquatic centres are complex building typologies. Every building is unique, and it is very difficult to find two aquatic centres with similar physical and operational characteristics. The high-energy consumption of aquatic and recreation centres creates both challenges and opportunities for energy conservation and the improvement of indoor environmental conditions. However, environmental design standards for aquatic centres have generally been overlooked due to the complex nature of these buildings. Consequently, the sector lacks both qualitative information and quantitative information and benchmarking guidelines. At present, there are no general Australian energy and water performance standards for public swimming pools, for buildings housing public pools or parts of buildings housing public pools [35]. The following section outlines some challenges faced in analysing the performance of aquatic centres.

2.1 *Inconsistencies in the Definition of Aquatic Centre*

A major issue around the aquatic and recreation industry sector is that there is not enough clarity regarding the definition of aquatic centres and the types of amenities they encompass. Aquatic centres have many functional areas. Swimming pools, sports halls and fitness centres cover the majority of floor areas in an aquatic centre. Duverge et al. [10] reviewed different terms and names that have been used in past studies to describe aquatic centres. They include aquatic leisure centres, public pools, aquatic and recreational centres, aquatic facilities, indoor swimming pools and leisure centres, public swimming bath, natatoriums, recreational facilities, sport facilities, sport complexes and leisure pool facilities. Also there have been changes in the statistical classification of the buildings that fall into the category of sport and recreation industry as per the Australian Bureau of Statistics (ABS) reporting [31]. Aquatic centres were included under health fitness centres and gyms in 2010, structured facilities such as gyms, public pools or courts in 2011 and outdoor sports facilities in 2013.

The study by Duverge et al. [10] emphasised how the lack of clear definition of aquatic centres can create confusion and difficulties when researching and comparing their energy and water usages. Duverge et al. [10] categorised aquatic centres according to the number and types of amenities that they offer and based on the data, defined an aquatic centre as a community or public venue that includes at least an indoor swimming pool and three different types of amenities such as a gymnasium, sauna/spa, café and crèche. Centres with only outdoor swimming pools are therefore not classified or defined as an aquatic centre. The majority of outdoor swimming pools in Victoria are usually only opened during the summer seasons. With the definition provided, it should be easier to distinguish between an aquatic centre and a small indoor swimming pool within a school as an example. Based on the definition, a review of all the facilities in Victoria was conducted. However, the number of aquatic centres was found to be different (110 numbers) from those recorded in the previous studies reported in the introduction section (500 numbers).

2.2 *Different Functional Areas and Indoor Environmental Quality Requirements*

Aquatic centres have many functional areas. Therefore, it is necessary to define a clear boundary of what is to be included in the energy analysis. Energy consumption of a centre with an indoor swimming pool is significantly higher than that of outdoor centres of the same size. However, recent studies have shown that to improve financial viability and produce higher participation rates, decision-makers involved in the planning of public aquatic centres should aim to include multi-purpose facilities with indoor swimming pools and minimise facilities with outdoor swimming pools [15]. Swimming pools have higher energy intensity than sports halls due to their specific requirements such as high latent and sensible heat loads, and ventilation loads. Sports halls generally have natural or hybrid ventilation systems, whereas gymnasiums and fitness centres are air-conditioned. In spaces such as stadiums and athletic halls, a large number of people may attend events and athletes may train and compete in a heavily polluted local environment [27]. A numerical study using computational fluid dynamics revealed that significant thermal stratification occurs in gymnasiums and that annual cooling loads can be overestimated by 45% if the effect of thermal stratification is not considered [18]. Rajagopalan and Luther [25] investigated the thermal and ventilation performance of a naturally ventilated (ceiling fan-assisted) sports hall using field measurements and numerical simulations. The measured results were analysed to develop various potential strategies for natural and low energy conditioning.

In the case of swimming pools, there has been a consistent trend towards higher water temperatures in recent years, due to the substantial growth in aquatic leisure activities [22]. A safe, comfortable and appealing internal environment is crucial to attract and sustain customers. According to American Society of Heating Refrigeration and Air-Conditioning Engineers (ASHRAE) Handbook [1], air temperatures in public and institutional pools should be maintained 1–2 °C above pool temperature to reduce the evaporation rate and avoid chill effects on swimmers. Ma et al. [19] also suggested that 1–2 °C is a suitable temperature difference between the water and the air. With an increasingly wide variety of pool uses, and flexible pool operations, it is difficult to select a single appropriate or optimum operating temperature for any particular pool [20]. Large amount of outside air is required to offset the amount of water in the atmosphere. Further, extensive water features will cause more evaporation and require more air change rates. According to Australian Standards 1668.2 [4] and pool operators' handbook [22], recommended indoor temperature for indoor swimming pool is 27 °C and relative humidity range is 50–60%. Higher temperature can cause discomfort to swimmers, thereby limiting vigorous swimming. Higher water and air temperature increase direct and indirect energy costs.

3 Proposed Methodology

A comprehensive methodology for evaluating the design and operational performance for aquatic centres is proposed in this chapter. Physical, operational and energy consumption data as well as environmental performance data are important information that needs to be collected for performance evaluation. Physical data about the building fabric and energy consuming equipment required to evaluate the design performance of the building can be obtained through design specifications and drawings. Energy consumption data required for evaluating the operational performance can be collected through utility bills and sub-meters. Installing sub-metering can be tedious and expensive due to the complexities in electrical switch board layout. Information about the physical characteristics such as floor area, water surface area and building envelopes can be obtained from architectural drawings and detailed site inspection.

Environmental performance such as thermal comfort and air quality measurements can be collected using various sensors that measure air temperature, mean radiant temperature, humidity and wind speed at the occupant levels. A movable thermal comfort cart will be appropriate for measuring comfort parameters at various locations. For identifying opportunities for improvement, the interrelationship between numerous factors that contribute to the energy consumption of these facilities should be understood and the significant drivers of building energy use can be determined.

3.1 Design Rating

Various systems that should be evaluated to determine the design rating include building fabric, air side mechanical, water side mechanical, air heating and water heating systems. These systems are discussed below in detail.

3.2 Building Fabric

In order to save energy, it is important that the building fabric is well constructed and sealed. Table 1 shows the details of the envelope construction and insulation of seven aquatic centres selected for a pilot study in Victoria. A relative ranking is assigned to each based on the construction details and possible implication on energy performance. The construction of old buildings and those constructed after 2006 varied considerably because of the Building Code of Australia's Section J requirements for the envelope [3]. Each building is assigned an energy efficiency rating according to their construction. Based on site inspection, new buildings are found to be well insulated, with double glazed windows, whereas some of the old buildings are found to be poorly insulated, with gaps in joints, window frames and mullions. It is important to make sure that uncontrolled air movement is minimised to prevent excess Heating Ventilation and Air Conditioning (HVAC) system work to treat the air.

Building envelope permeability rate, expressed in cubic metres of air per hour that passes through each square metre of the building façade ($m^3/h/m^2$), is a measure of the amount of air that permeates through gaps in the building envelope when driven by external forces such as wind or the thermal stack effect and can be tested using air pressure testing. Table 2 shows different permeability rates as specified by the Air Tightness Testing and Measurement Association [2]. Previous studies have shown that air leakage rates in Australia are much higher than those reported in Europe and USA [12]. From Table 1, it can be seen that Building 3 constructed in 2014 has air permeability rating of 10 $m^3/h/m^2$. Building 5 with medium age has a permeability rating of 22.5 $m^3/h/m^2$. Reducing leakage from 22.5 $m^3/h/m^2$ to below 15 $m^3/h/m^2$ requires careful sealing of all air gaps.

Table 1 Type of building fabric

Facilities	Age of buildings (years)	Building fabric	R value (m^2K/W)/SHGC	Energy efficiency rankings	Permeability rating ($m^3/h/m^2$)
Building 1	20	Insulated walls with no sealing, single glazed windows with air gaps in mullions/window frames	1–1.2 for walls, 0.9 for windows/SHGC 0.8	4	Not measured
Building 2	35	Minimal insulation, with air gaps, single glazed windows with air gaps in mullions/window frames	1–1.2 for walls, 0.9 for windows/SHGC 0.8	5	Not measured
Building 3	1	Part J compliant, well-sealed walls and roofs, double glazed windows	1.8–2.8 for walls, 1.6–1.8 for windows/SHGC 0.6	1	10
Building 4	11	Insulated walls and no sealing, single glazed windows with air gaps in mullions/window frames	1–1.2 for walls, 0.9 for windows/SHGC 0.8	4	Not measured
Building 5	18	Insulated walls reasonable sealing single glazed and sealed windows	1.2–1.5 for walls, 0.9–1.2 for windows/SHGC 0.8	3	22.5
Building 6	3	Part J compliant, well-sealed walls and roofs, double glazed	1.8–2.8 for walls, 1.6–1.8 for windows/SHGC 0.6	2	Not measured
Building 7	36	Minimal insulation, with air gaps, single glazed windows with air gaps in mullions/window frames	1–1.2 for walls, 0.9 for windows/SHGC 0.8	4	Not measured

Table 2 Permeability rating as per ATTMA Technical Standard 2

Type of facilities	Air permeability (best practice)	Air permeability (good practice)
New leisure and aquatic centres	2	5
20 year+ old leisure and aquatic centres	5	15

Table 3 Design rating for air-side mechanical system

Fans	Heat recovery	Fresh air percentage	
All variable speed fans	Full heat recovery heat wheel	100 per cent fresh air with heat recovery	Best practice ↑
No variable speed fans	Run-around heat coil	100 per cent fresh air with no heat recovery	↕
	No heat recovery	50:50 recycled air	↓
		100 per cent recycled air	Worst practice

3.3 Air Side Mechanical System

Proper air distribution is also important in addition to how much outside air is introduced. Airflow over the pool surface and deck area should be minimised to reduce drafts on swimmers and the rate of evaporation, as draft increases with air velocity. Lack of proper air distribution can sometimes be monitored through evidence of condensation and can be minimised by directing air supply onto indoor window surfaces. Mechanical air handling equipment should be capable of 100% fresh air supply to provide, either through energy recovery or through economy-cycle options, the largest amount of fresh air possible while maintaining thermal comfort. Variable air volume (VAV) systems can manage changing load requirements by varying the airflow to keep the temperature constant in different thermal zones. The advantages of VAV systems over constant-volume systems include more precise temperature control in thermal zones, lower fan energy consumption and reduced fan noise. There is considerable scope for the use of heat recovery systems in pool ventilation.

The most important means of heat recovery in a pool is often sensible heat recovery. There are three main types of heat recovery systems [8]: cross flow heat exchangers, run-around coils and thermal enthalpy wheel system. Previously regarded as a high-cost specialist heating component for very large space applications, the thermal enthalpy wheel is now seen as a highly economical form of energy saving and is rapidly becoming popular in new buildings where large volumes of air need to be handled and high efficiency is required. The pilot study helped to consolidate some of the main considerations for identifying the best practice and ranking the design of air-side mechanical system as indicated in Table 3.

3.4 Water Side Mechanical System

Water filtration is very important to pool water treatment at aquatic centres. The main components of water treatment include circulation pumps, heat recovery, filtration and backwash, ultraviolet treatment plant, water disinfection and total dissolved

Table 4 Design rating for water-side mechanical system

Circulation pumps	Heat recovery	UV treatment plant	
VSD ramp speed reduced by 20–30 per cent based on turbidity, Nephelometric Turbidity Unit (NTU), provided chlorine, chloramines and pH levels are within the limits	Heat recovery	Auto power control based on chloramines with override if chlorine is below set point	Best practice
VSD fitted, but speed fixed to achieve constant turnover rate when open (timer controlled for after-hours slow down)		Auto power control based on stepless UV level	
VSD fitted, but speed fixed to achieve arbitrary and constant turnover rate		Auto power control based on stepped UV level	
VSD not fitted or fitted and not used for speed control		Manual lamp power control	
	No heat recovery	No lamp power control	Worst practice

solids. Maintaining the quality of swimming pool water requires pumps that consume electricity. Swimming pool water is continually circulated through filters to capture contaminants. Unnecessary filter backwashing wastes water, energy and chemicals due to the need to heat and treat the incoming make-up water. The frequency and volume of the water used in each backwashing cycle depend on the filter type, filter media and operation of the filters during backwash [29]. For convenience and simplicity, filters are often backwashed to a schedule for a set period. Best practice filter operation is to backwash only as necessary.

Water quality is the most important aspect of an aquatic centre's operation. Clean, clear and healthy water attracts bathers and ensures bather safety [29]. Chlorination is widely used to disinfect pool water. It is generally achieved by adding sodium or calcium hypochlorite to the water. Relatively large volumes of sodium hypochlorite are needed to maintain water quality and a sufficient residual. However, chlorination can quickly lead to the build-up of total dissolved solids (TDS). Compared to sodium hypochlorite, a smaller proportion of calcium hypochlorite is required. Thus, it builds up more slowly and less diluted water is required to mitigate TDS. Some aquatic centres supplement chlorination with ultraviolet (UV) light irradiation or ozone treatment for extra protection and to reduce chemical use. UV and ozone treatment have no effect on pH or water balance and do not contribute to TDS levels. However, they do help to reduce volatile gases and smell. Ozone plants may have high on-going operational and maintenance costs. Table 4 shows the main considerations for ranking the performance of water-side mechanical system based on the pilot study.

3.5 Heating

There are many issues to consider when selecting a pool heating system, including capital and running costs, fuel tariffs, the space allocated for the equipment, the location of heating equipment, the availability of energy, energy costs and any budgetary restraints (e.g. capital and operating budgets).

The method used to heat pool water varies from centre to centre and is an important consideration for comparing buildings. The type of fuel used depends on the councils' environmental policies, existing contracts with the energy providers and price of gas and electricity. The most common methods of producing heat include cogeneration units, condensing boilers, standard boilers and heat pumps. The seasonal efficiencies of these systems vary. Condensing boilers are more effective than standard boilers. Heat pumps are probably most cost-effective as part of a heat recovery/dehumidification system. Cogeneration system will significantly change the energy and emissions profile of the facilities and councils. Pools utilising cogeneration aim to use more gas but less electricity resulting in fewer greenhouse gas emissions. It is to be noted that savings largely depend on the size and effectiveness of the cogeneration system and the base electricity load. This is due to cogeneration plants requiring to operate ideally at clearly defined steps (e.g. 50, 100% load). In order to achieve the maximum efficiency of cogeneration units, they need to run at full load for as many hours as possible.

A significant proportion of the total energy consumption is in the form of pool water heating. Evaporation of pool water is the major source of heat loss. Covering a swimming pool when it is not in use is the most cost-effective way of reducing pool heating. A swimming pool cover limits the exposure of the swimming pool surface to the surrounding air by providing a physical barrier between the swimming pool surface and the atmosphere. The set point temperature can be reduced when the swimming pool is covered, and this allows a significant amount of energy to be saved. Properly designed swimming pool covers have the following benefits:

- Reduce water losses by 30–50%;
- Reduce heat losses by 70–90%;
- Reduce HVAC running costs; and
- Reduce the effect of condensation on the building structure, fabric and fittings [29].

Heat from the pool water is also lost when backwashing of filters or when make-up water is added. Heat recovery is possible from the backwash using heat exchangers. Some of the facilities use solar thermal to complement the heating systems. A recent study [13] noted that there has been only a marginal increase in uptake of these types of systems in municipal pools in Victoria compared to 30 years ago and the main perceived barriers to the installation of solar systems are similar to those of 30 years ago. These misconceptions are cost, lack of roof area and inability of solar to meet the energy needs of aquatic centres [13]. Table 5 shows the various types of pool heating system and possible ranking based on the pilot studies.

Table 5 Design rating for heating system

Swimming pool heating	Swimming pool covers	
Condensing boiler with cogeneration + solar	Automatic swimming pool cover or swimming pools with Bauer control system	Best practice
Condensing boiler with cogeneration	Manual swimming pool cover	
Condensing boiler with solar	No swimming pool cover	
Heat pumps with solar		
Condensing boiler		
Heat pumps		
Boiler—standard boiler with solar		
Boiler—standard boiler		
Direct pool boiler		Worst practice

3.6 *Operational Rating*

Operational performance is evaluated by comparing the actual energy consumption and usage intensity with previously developed benchmarks. Aquatic facilities, like most other building types in Australia, use both electricity and gas. Electricity is used for lighting, gym equipment, pumps, fans, etc., while gas is used for space and pool heating. Motors, fans and pumps were used widely throughout the swimming pool buildings for water treatment and ventilation systems. Energy use per unit area or energy usage intensity (EUI) is the most commonly used indicator in benchmarking studies. However, a lack of clarity exists as to how to determine the best indicators. Some studies have used usable area [7, 28] and water surface area [17, 32] as performance indicators.

Using water surface as a performance indicator makes energy comparisons between aquatic centres and other types of building difficult. As a performance indicator, water surface area might be appropriate if only indoor swimming halls are being considered in the benchmarking. However, most aquatic centres include several amenities and dry areas (e.g. gymnasiums, sport halls and cafés). The number of visitors also has an impact on the energy consumption of centres. A study involving six aquatic centres in Victoria showed that the EUI per floor area values ranged from 1824 to 5983 MJ/m^2 [24]. Here gross floor area included both conditioned and unconditioned areas. The unconditioned floor area such as storerooms, plant rooms, multi-purpose sports halls and basketball stadiums must be excluded from the total floor area unless they used a significant amount of energy or water. As an example, an unconditioned sport hall should not be included if possible because including the area of a sport hall (they usually have a large floor area) in a benchmark analysis will lower the centre's energy intensity compared to a centre that does not have a sport hall. The calculated electricity of sport hall lighting could be deducted from the overall electricity usage.

Normalising with the number of visits showed that there was a considerable variation among the values which ranged from 25 to 76 MJ/visit (Rajagopalan 2014). Visitors normally use the dry amenities such as gymnasiums, sport halls and childcare facilities and the wet facilities such as swimming pools. Visitors that use swimming pools contribute to the bather load and will use more energy and water compared to visitors using dry amenities. However, most of the centres do not separately record the number of visitors using the swimming pool; thus, the total number of visitors had to be used in the analysis. The number of annual visitors at the six centres ranged from 168,000 to 1,200,000 people. Duverge et al. [10] suggested that a possible method for calculating the number of persons could be that one bather will be recorded as one person. Every 5.5 persons utilising the dry area amenities or visiting the centre will be accounted as one bather because bathers will use more water than people using dry area facilities or visiting the centre.

4 Indoor Environmental Performance

Operational energy performance should be evaluated in conjunction with indoor environmental quality to make sure that energy efficiency is achieved not by compromising indoor environmental quality. The main indoor environmental quality measures that have direct implications on energy include thermal comfort, indoor air quality and lighting.

4.1 Thermal Comfort

There are limited studies comparing the thermal comfort conditions of indoor aquatic facilities. Even though several thermal comfort studies are conducted in other building types, such studies are limited with respect to aquatic centres. With the existence of multiple user groups, achieving thermal comfort has always been challenging. The suitability of available thermal comfort models are not tested for these building types. In a recent study, thermal comfort conditions of various user groups in seven aquatic centres in Australia were investigated during the winter period in the months of July to August (outdoor temperature ranged between 5 and 16 °C), in conjunction with monitoring environmental parameters and surveying various user categories [26]. In this study, thermal comfort evaluation of the pool environment through the use of PMV and thermal vote (TV) was performed for three user groups: swimmers, spectators who care for children undertaking swimming lessons and staff members who work as swimmer supervisors were taken into consideration.

The set point temperatures measured ranged between 24 and 32 °C and found to be significantly high in some of the buildings resulting in high level of discomfort for the spectators and staff. The air temperatures were found to be lower than water temperatures in most of the facilities investigated. Considering the average external and internal temperatures, significant amount of energy is required to heat the air from 10 to 29 °C and to keep the water also at similar temperatures. However, evaporation loss will be more if the air temperature is set low.

The study found that thermal vote for various user groups was significantly different from each other in most of the facilities. Among the three groups of users, only the thermal sensation of staff was significantly correlated with the calculated PMV. In addition, the thermal sensation of spectators indicated correlation with PMV. As the PMV model cannot be used for activity levels more than 2.0 Met, it has limitations in terms of metabolic rate in predicting the comfort of swimmers. In a previous study, PMV model was applied for swimmers without any specific correction after 10-min post-swimming [20] However, for wet simmers, the heat balance for the evaluation of the PMV requires the addition of the evaporative term.

4.2 Air Quality

Poor air quality in an indoor swimming pool centre can have a negative effect on the health of swimmers, coaches and swimming pool workers and can lead to respiratory problems. Some of the parameters that influence the air quality such as carbon dioxide and chloramine levels are discussed here.

4.2.1 Carbon Dioxide Levels

CO_2 levels below 1000 ppm in indoor environments indicate adequate air circulation. CO_2 concentrations in outdoor air typically range from 400 to 600 ppm. Figure 1 shows the CO_2 levels measured over the two-day period in six aquatic centres in Victoria. Building 1 and Building 2 had higher levels of CO_2 with concentrations increasing up to 1600 ppm during certain periods. Most of the facilities used carbon dioxide gas to reduce the pH level of water. Uncontrolled CO_2 use in the water resulted in high CO_2 levels in the pool hall air for some facilities. Building 6 had the lowest level of CO_2 (<600 ppm) for the entire measurement period. Using specialised energy management control system such as Bauer system seemed to work well particularly in a Building 3. This system improves the mixing of the air at a molecular level and significantly decreases the effect of inversion layers within the space. The effect of an inversion layer is a significant temperature difference between the top and bottom of an expansive space like that commonly found in indoor aquatic centres. The system creates a very small positive pressure in the space by controlling the speed of the supply fans and controlling the position of the outside air dampers within the system. The slightly positive pressure then allows the air to mix at a partial pressure level. Conventional HVAC results in temperature stratification within occupied zone where hot air rises towards the roof and cold air descends. Bauer system monitors zone and duct pressure, temperature, humidity, IAQ and/or CO_2 levels, supply and return

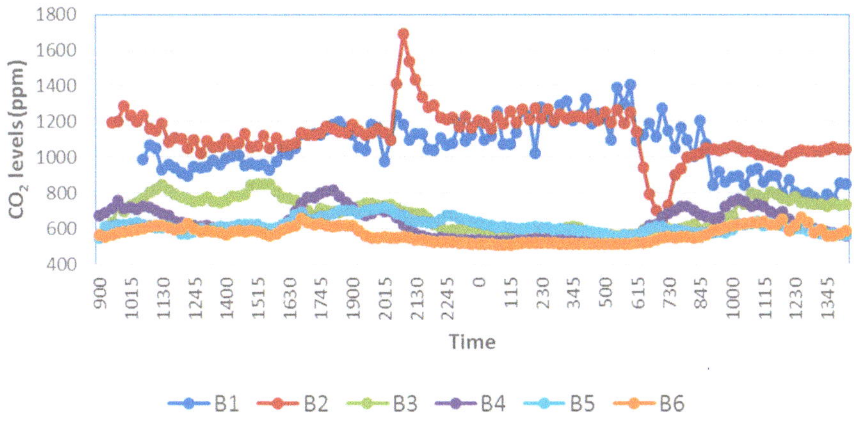

Fig. 1 Carbon dioxide levels measured in the six aquatic centres

damper positions to determine the volume and quality of air needed and typically operates a lower supply air velocities, hence saving fan energy.

4.2.2 Chloramines

Aquatic centres require considerable disinfection to avoid microbiological pollution; however, paradoxically, disinfection by-products can create health hazards. Individuals commonly walk into indoor swimming pool facilities and perceive a smell of 'chlorine', but this odour is not caused by excess chlorine; rather it is caused by a chlorine compound called chloramine that is formed in the water and released from the surface of the pool. Several studies have shown that associated increases in asthma and respiratory illnesses in swimming pool patrons, lifeguards, coaches and observers are the result of exposure to disinfection by-products. Further, these by-products are also the primary cause of facility corrosion.

Over the past few years, research has focused on the quality of pool hall air quality and, in particular, the effects of chronic lung exposure to chlorine and its by-products, especially in young children [6]. Respiratory symptoms and asthma is more prevalent in competitive swimmers than other athletes [14]. Thickett et al. [30] found that trichloramine can be a cause of occupational asthma in swimming instructors and lifeguards. A study of Jacobs et al. [16] showed an elevated prevalence of respiratory symptoms in Deutsch swimming pool workers compared to the general population. Parrat et al. [21] showed that even relatively low exposures to trichloramine (i.e. up to 0.3 mg/m^3) could cause health problems. Predieri and Giacobazzi [23] noted that the most sensitive populations to environmental factors are babies and young children, as at these ages, organisms are more vulnerable to toxins because immunological and lung development is not yet complete.

The quantity of trichloramine emitted from pool water depends on various factors such as water temperature, water agitation, number of swimmers, the concentration of urea and free chlorine as well as air ventilation conditions. However, the chemical composition of swimming pool air is extremely complex, and how these factors are associated with trichloramine levels in the air is poorly understood [33]. Due to energy costs, the fresh air ratio is often reduced, leading to insufficient by-product reduction [21]. Air circulation systems should be able to distribute air effectively over the whole of the pool hall area to eliminate any chlorine odours, risk of condensation and uncomfortable drafts. As a result of a few studies conducted in the late 1990s, the World Health Organization [34] recommended a reference value of 0.5 mg/m^3 of trichloramine in the air.

5 Discussions

Aquatic centres use vast amount of energy and water. The increasing use of indoor aquatic centres results in high energy consumption which in turn gives rise to a large carbon footprint for this building type. However, environmental design standards for aquatic centres have generally been overlooked due to the complex nature of these buildings. Through analysis of energy consumption data, various normalisation factors were identified. The factors include total floor area, pool surface area and total number of visitors. Selecting the most suitable indicator that can accurately predict the energy consumption is very challenging and requires rigorous statistical analysis involving a large number of samples.

Measurements conducted in selected centres in winter showed that comfort experiences vary from one building to another. Achieving thermal comfort is challenging, especially with the multiple user groups. The set point temperatures are significantly high in some of the buildings studied and this resulted in high level of discomfort for the spectators and staff. Spectators tend to wear more clothes in cool days as they are free to choose the clothes they wish to wear and adapt to the environment. Clothing adaptation is one way of improving comfort conditions for spectators. It is also important to provide comfortable staff uniforms. The temperature settings should be changed according to the seasons rather than keeping constant throughout the whole year. This will help to improve the comfort in relation to the adaptation and expectation of the occupants. There are limited studies comparing the thermal comfort conditions of indoor aquatic facilities. The suitability of available thermal comfort models are not tested for a range of indoor aquatic centres.

Smell of chlorine, high humidity and carbon dioxide levels are common in aquatic centres. Any air circulation system should be able to distribute air effectively over the whole of the pool hall area to eliminate any chlorine odours, the risk of condensation and uncomfortable drafts. Further, condensation on large glazed surfaces is not uncommon and promotes the growth of mould. Despite the considerably different heating ventilation and air-conditioning systems used in Australia, to date, no studies have been published on the level of trichloramines in the Australian aquatic centres. A proper balance of chloramine control, air distribution, outdoor air and room exhaust air along with air movement at the water surface is crucial to ensure good indoor air quality. Chloramines also corrode handrails, ladders, exposed steel structural elements and HVAC components. It is to be noted that trichloramine measurement is an expensive process and involves collecting air samples in filters through sampling pumps and performing chromatography analysis. As atomic weight of NCl_3 is much heavier than oxygen, there is more prevalence of trichloramine at low levels in the pool halls. Locations for sampling should be carefully selected and focused on areas where atmospheric concentrations are likely to be high, for example areas close to return air and areas of water agitation.

Each of the indoor environmental parameters has implication on the energy consumption of the building, particularly heating energy. The relative influence of various physical and operational parameters on the total energy consumption can be investigated using energy simulations. Modelling the whole aquatic centre accurately, particularly the pool hall, can be challenging. Energy required for pool water heating depends on the evaporation rates. Management of the rate of evaporation is an important aspect of indoor pool environment care. Higher relative humidity generally results in lower energy use, but must not be so high as to cause indoor air problems or structural damage.

6 Conclusions

This chapter provided insights into the complexity of aquatic centres and their variability in terms of energy performance. Also a methodology for assessing their design and operational performance is developed. One of the main indicators for benchmarking operational performance is identified as conditioned floor area. Detailed measurements conducted in some of the Victorian facilities showed that indoor parameters across the buildings vary significantly as a result of which comfort experiences differ from one building to another. The majority of the buildings constructed before the year 2006 had porous facades resulting in energy wastage through unwanted infiltration and exfiltration. The impact of building envelope and uncontrolled leakage on energy consumption for these building typologies need to be further investigated through energy simulation. This methodology will assist local government to manage energy consumption, indoor environmental quality and system design through better informed day-to-day management of operations, as well as guiding environmental performance during the design of new or refurbished infrastructure.

References

1. ASHRAE, HVAC Applications (2016) American Society of Heating, Refrigerating and Air-Conditioning Engineers Inc., Atlanta, GA, United States

2. ATTMA (2015) Measuring air permeability of building envelopes, the air tightness testing and measurement association, c/o the British institute of non-destructive testing, technical standard L2, measuring air permeability of building envelopes https://www.attma.org/wp-content/uploads/2016/04/ATTMA-TSL2.pdf. Last accessed 20 Jan 2018
3. Australian Building Codes Board (2016) National construction code volume 1 energy efficiency provisions handbook. ACT, Canberra
4. Australian Standards 1668.2 (2012) The use of mechanical ventilation in buildings. Standards Australia, Sydney, NSW
5. Australian Water Safety Council (2007) Public swimming pool legislation in Australia: an examination of the legislation. Available from https://www.royallifesaving.com.au/__data/assets/pdf_file/0010/4024/Final_Position_Paper_Public_Swimming_Pool_Legislation_Dec_2007.pdf). Accessed 10 Oct 2017
6. Bernard A, Carbonnelle S, Michel O, Higuet O, de Burbure C, Buchet JP, Hermans C, Dumont X, Doyle I (2003) Lung hyperpermeability and asthma prevalence in schoolchildren: unexpected associations with the attendance at indoor chlorinated swimming pools. Occup Environ Med 60(6):385–394
7. British Swimming (2008) The use of energy in swimming pools. Available from http://www.swimming.org/assets/uploads/library/Use_of_Energy_in_Pools.pdf. Accessed 26 Aug 2014
8. Building Research Energy Conservation Support Unit (BRECSU) (2001) Energy consumption guide, 78, Energy use in sports and recreation buildings. Available https://www.cibse.org/getmedia/34def23a-c65b-405e-9dff-ce181c0b1e0d/ECG78-Energy-Use-in-Sports-and-Recreation-Buildings.pdf.aspx. Last accessed 10 Oct 2017
9. Carbon Trust (2006) Sports and leisure introducing energy saving opportunities for business. https://www.carbontrust.com/media/39352/ctv006_sports_and_leisure_sector_overview.pdf. Last accessed 13 Jan 2018
10. Duverge J, Rajagopalan P, Fuller R (2017) Defining aquatic centres for energy and water benchmarking purposes. Sustain Cities and Soc 31:51–61
11. European Commission (2015) 2030 framework for climate and energy policies. Available from https://ec.europa.eu/clima/policies/strategies/2030_en. Accessed 8 Jan 2018
12. Egan AM (2011) Air tightness of Australian offices buildings: reality versus typical assumptions used in energy performance simulation. In: Proceedings of the 12th conference of international building performance simulation association, Sydney
13. Fuller R, Rajagopalan P, Duverge JJ (2017) Assessment and modelling of the viability of a solar heating system for aquatic centres in southern Australia. Energ Effi 10(5):1269–1278
14. Goodman M, Hays S (2008) Asthma and swimming: a meta-analysis. J Asthma 45:639–647
15. Howat G, Murray D, Crilley G (2005) Reducing measurement overload: rationalizing performance measures for public aquatic centres in Australia. Managing Leisure 10:128–142
16. Jacobs JH, Spaan S, van Rooy GBGJ, Meliefste C, Zaat VAC, Rooyackers JM, Heederik D (2007) Exposure to trichloramine and respiratory symptoms in indoor swimming pool workers. Eur Respir J 29:690–698
17. Kampel W, Aas B, Bruland A (2013) Energy use in Norwegian swimming halls. Energy Build 59:181–186
18. Lam CJ, Chan LSA (2001) CFD analysis and energy simulation of a gymnasium. Build Environ 36(3):351–358
19. Ma X, Jian Y, Cao Y (2006) A new national design code for indoor air environment of sports buildings. Facilities 2(11/12):458–464
20. Revel Marco (2014) Arnesano, Perception of the thermal environment in sports facilities through subjective approach. Build Environ 77:12–19
21. Parrat J, Donze G, Iseli C, Perret D, Tomicic C, Schenk O (2012) Assessment of occupational and public exposure to trichloramine in Swiss indoor swimming pools: a proposal for an occupational exposure limit. Ann Occup Hyg 56(3):264–277
22. Pool Operators' Handbook (2008) Victorian Government Publishing Service, Victoria, Australia

23. Predieri G, Giacobazzi P (2012) Determination of nitrogen trichloride (NCl3) levels in the air of indoor chlorinated swimming pools: an impinger method proposal. Int J Environ Anal Chem 92(6):645–654
24. Rajagopalan P (2014) Energy performance of aquatic facilities in Victoria, Australia. Facilities 32(9/10):565–580
25. Rajagopalan P, Luther M (2013) Thermal and ventilation performance of a naturally ventilated sports hall within an aquatic centre. Energy Build 58:111–112
26. Rajagopalan P, Jamei E (2015) Thermal comfort of multiple user groups in indoor aquatic centres. Energy Build 105:129–138
27. Stathopoulou OI, Assimakopoulos VD, Flocas HA, Helmis CG (2008) An experimental study of air quality inside large athletic halls. Build Environ 43(5):834–884
28. Step2Sport (2015) Step by step renovation towards nearly zero energy sport building. http://step2sport.eu/?page_id=8. Accessed 9 July 2017
29. Sydney Water Corporation (2011) Best practice guidelines for water management in aquatic leisure centres. Available at https://www.sydneywater.com.au/web/groups/publicwebcontent/documents/document/zgrf/mdq1/~edisp/dd_045262.pdf. Last accessed 25 Aug 2017
30. Thickett KM, McCoach JS, Gerber JM (2002) Occupational asthma caused by chloramines in indoor swimming pool air. Eur Respir J 19:827–832
31. Tower J, McDonald K, Steward B (2014) Community benefits of Victorian aquatic and recreation centres: technical report for aquatic and recreation Victoria. Victoria University, Melbourne, Australia
32. Trianti-Stourna E, Spiropoulou K, Theofilaktos C, Droutsa K, Balaras CA, Santamouris M (1998) Energy conservation strategies for sports centers: part B: swimming pools. Energ Build 27(2):123–135
33. World Health Organization (WHO) (2000) Guidelines for safe recreational water environment. Volume 2: swimming pools, spas and similar recreational water environments. WHO, Geneva, Switzerland
34. WHO (2006) Guidelines for safe recreational water environments, Volume 2, swimming pools and similar environments. WHO, Geneva, Switzerland
35. Wilkenfeld G (2009) Swimming pools and electric space heating: The case for coverage by the Building Code of Australia. Retrieved 21 Aug 2014. www.abcb.gov.au

Assoc. Prof. Priyadarsini Rajagopalan, B.Tech. Arch. Eng., M.Bld.Sc., Ph.D. is an associate professor with expertise in building energy efficiency and energy benchmarking, indoor environmental performance and urban microclimate.

A Feasibility Study and Assessment: Distributed Solar System in High-Density Areas

Rebecca J. Yang and Andrew Carre

Abstract Solar photovoltaic (PV) energy has emerged as an innovation for greenhouse gas reduction in the building and construction industry due to the calculable advantages it possesses. Although there is evidence supporting the inclusion of small-scale roof-mounted PV systems in detached houses, limited studies have been conducted on the implementation of PV in the commercial sector especially within high-density urban areas. This study conducted a detailed value assessment to optimize the cost of applying PV systems in a high-density city area of Melbourne. The Net Present Value results evidence the feasibility to apply roof-mounted polycrystalline PV products in the case study buildings. This research supports investors' decisions by understanding the financial values of prefabricated PV systems in high-density regions and provides suggestions to building professionals on value-for-money design.

1 Introduction

Established buildings are responsible for 32% of the average total energy consumption globally (IPCC, [14]; see also Chapter "The built environment in Australia"). From this global perspective, governments are committed to making nonrenewable energy sources redundant. Solar photovoltaic (PV) energy has emerged as an alternative renewable energy source in the building and construction industry due to the calculable advantages it possesses. Australia has high potential for solar energy compared to many regions around the world due to the high levels of solar radiation discharged (Department of Industry, Innovation and Science, 2016). Yet, Australia is also one of the leading coal-burning countries in the world [9]. Coal power stations in Australia are producing 170 million tons of carbon dioxide annually, contributing to climate change [5]. If the right policies are introduced, it is expected that consumers

R. J. Yang (✉) · A. Carre
Sustainable Building Innovation Laboratory, School of Property Construction and Project Management, RMIT University, Melbourne, VIC, Australia
e-mail: rebecca.yang@rmit.edu.au

P. Rajagopalan et al. (eds.), *Energy Performance in the Australian Built Environment*, Green Energy and Technology, https://doi.org/10.1007/978-981-10-7880-4_11

using solar panels and batteries will produce between 30 and 50% of Australia's electricity needs by 2050 [10]. Furthermore, the Climate Commission report [7] predicts solar technologies to improve and have higher efficiency and lower cost in the near future, which in turn will tap into the widespread market.

In line with the Paris Agreement, the Australian government has set a renewable energy target of delivering 23% of Australia's electricity from renewable sources by 2020 [8], with expected increases of large-scale PV products focusing on buildings due for renovations and new developments, forming part of achieving this target. In Australia, more than 20% of residential buildings have installed a solar PV system and more than 80% of these are grid-connected [1]. Attractive government policies and other incentive programs have resulted in steady growth of PV adoption. Although, there is firm growth of rack-mounted PV systems in the residential buildings, especially detached houses [1], limited research has been conducted in Australia on the values assessment of PV applications in the commercial sector especially within high-density urban areas. Moreover, with the rapid population growth and large number of developments in major Australian cities such as Sydney and Melbourne, significant increase in energy demand is anticipated in the central business district (CBD) areas. A feasibility study on the application of PV systems in these areas would provide insights for both public and private sectors on future incentives and practices to encourage the uptake of solar panels.

The design and development of solar energy systems require the knowledge of variation and maximum utilization of solar radiation the system is exposed to [22]. One of the main concerns of installing a photovoltaic system is achieving the maximum energy output and avoiding shading. In this study, a detailed value assessment is conducted to optimize the cost of applying PV systems in a high-density city area of Melbourne. This chapter reports on the development of two PV design scenarios for a case study of ten buildings located in Melbourne CBD. The design elements in the application of PV systems in buildings and the case study results will be discussed.

2 A Framework for Feasibility Study

This chapter reports on the results of a comprehensive study on analyzing the value of applying solar panels to Melbourne's CBD commercial buildings. According to the Melbourne City Council, the number of weekday daily workers in the city has grown over the past couple of years. It is estimated that more than 386,000 people travel to the city to work, or undertake work-related activities, on an average weekday [6]. Along with overall employment growth, supply of office floor space in the city grew substantially over the past two years. The City of Melbourne's development activity monitoring report suggests that more than 217,000 m^2 of office floor space was constructed in the city in 2013 and 2014. There are currently 20,538 private dwellings in Melbourne (CBD). By 2036, this is expected to increase to 49,868 [6]. All of these evidences suggest a high future energy demand in the CBD area. The inclusion of renewable energy, and specifically PV, can help to reduce this.

There are a number of factors that need to be considered to produce the most appropriate design to maximize PV system output. The potential site needs to be investigated, and the adjoining building heights or vacant blocks need to be factored into performance analysis. Partial shading is a major challenge imposed by taller neighboring buildings. Shading is also a major future risk, as cities grow and the buildings get taller the likelihood of overshadowing buildings is high which will result in energy losses. Shading challenges could be the difference between a viable and nonviable PV project. The tilt angle of a solar PV system influences the amount of energy collected by a PV module [15]. According to Gregg et al. [11], the primary reference point for determining the tilt angle is the latitude. They have further explained that the arc of the sun varies with the time of year, so the shallow tilt angles appear to produce more energy in the summer months, while the steeper tilt angles are more efficient in the winter months. There are many studies that investigate the optimal tilt angle of solar PV in various locations and situations, for example, East-Central Europe—Jantsch et al. [15]; Saudi Arabia—Benghanem [4]; Egypt—Hussein et al. [12]; Turkey—Bakirci [3]; and Canada—Rowlands [19]. These studies concluded that the optimal tilt angle is the angle at which the greatest amount of solar radiation will be received over the year and comparison study should be conducted to identify the best angle.

To determine the tilt angle of optimal PV orientation, the environment around the site should be inspected. Tall or large structures, high-rise buildings, trees, and surrounding site may create shadows onto the tilted solar panels throughout the day or when sun has a low angle during the winter. Hence, depending on their location on the building, installing flat solar panels (i.e., horizontal and facing up to the sky) might be the best way for PV systems to be installed as it allows each of the panels to receive maximum amounts of solar radiation at all times. On the other hand, when the site has no surrounding obstructions to influence the PV areas, in this case, a tilted design panel can be added to the PV system installation. In order to discover the best angle for PV, the National Renewable Energy Laboratory (NREL) [16] p. 8 states that "the PV system should always face toward the equator. In the Northern Hemisphere the panel should face south and tilt from horizontal at an angle approximately equal to the site's latitude." For example, if the PV system is placed in Melbourne, Australia, the solar panels should face north and have a tilt angle of approximately 30°–37°. Therefore, two design options are proposed in this study, namely flat (horizontal layout) and 30° north-facing PV designs.

A case study is the primary method used in this study. Five groups of ten buildings located in Melbourne CBD were selected to develop a case study to understand the cost–benefit of applying PV systems on building roofs. Buildings that were connected to the same electricity meter were grouped together. Table 1 shows the details of the five groups. Multiple data sources and tools were used in this study to conduct the feasibility study as shown in Fig. 1. (1) The original structural and architectural drawings were collected from the project client. (2) A third-party PV design firm, which has more than ten-year experience in PV design and construction, was involved to develop the alternative PV designs by using *PVSyst* tool [18]. This brings professional industry knowledge to the designs and, most importantly, creates

Table 1 PV Performance ratio on energy generation

Month	Building group 1 (Two Buildings)	Building group 2 (One Building)	Building group 3 (Three Buildings)	Building group 4 (Three Buildings)	Building group 5 (One Building)
Jan	0.50	0.50	0.77	0.70	0.50
Feb	0.76	0.83	0.83	0.83	0.69
Mar	0.75	0.90	0.85	0.75	0.72
Apr	0.77	0.83	0.77	0.73	0.74
May	0.71	0.84	0.71	0.67	0.68
Jun	0.73	0.90	0.76	0.71	0.76
Jul	0.71	0.52	0.69	0.63	0.62
Aug	0.77	0.90	0.77	0.69	0.74
Sep	0.77	0.90	0.84	0.72	0.73
Oct	0.76	0.83	0.88	0.83	0.73
Nov	0.76	0.83	0.88	0.85	0.73
Dec	0.77	0.83	0.88	0.86	0.73

the feasible designs from client's perspective. (3) After the alternative designs were developed, PV suppliers and installers were approached to obtain the capital cost information. A local utility provider advised the electricity price and feed-in tariff. Government incentive information was also obtained from the local council. (4) The actual hourly electricity consumption data was retrieved from the project client. (5) NREL's *PVWatts Calculator* [17] is a popular tool to predict energy outputs from solar panels. *PVWatts* provides the solar irradiation data and was used for comparison. A 3D model of the building was created using *IES-VE Suncast* [13] to simulate the shading impact. (6) Energy outputs were calculated based on the solar irradiation, system efficacy and loss. Energy consumption and output data were compared to identify the possible energy export to the public grid. (7) The Net Present Value (NPV) method and sensitivity analysis were conducted to show the cost–benefits of the alternative designs. The following section explains the method further and detailed case analysis outcomes.

3 Case Study Results and Inspiring Observations

3.1 Shading Impact Analysis and Energy Generation Performance Ratio

IES-VE Suncast was used to simulate the average monthly solar irradiation values of the case buildings (as shown in Fig. 2). These values were used to calculate

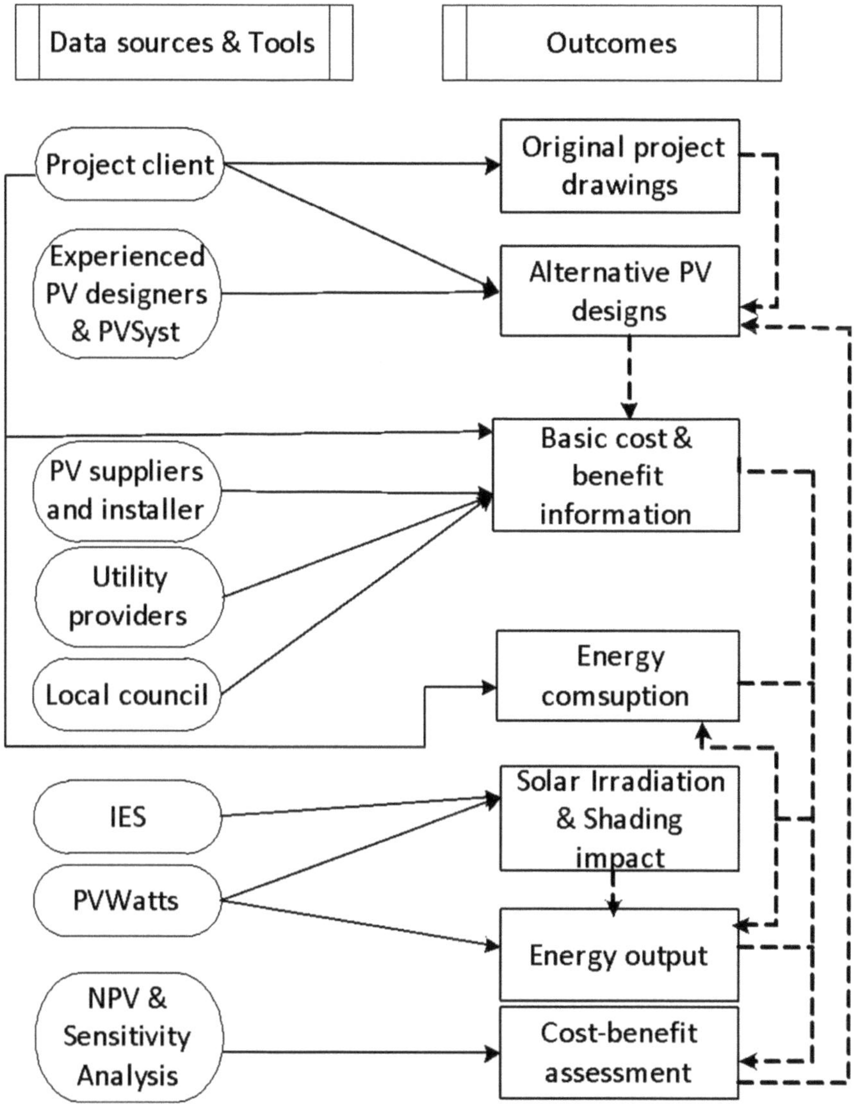

Fig. 1 A framework for PV value assessment

the monthly shading loss of each building group. The total performance loss was calculated by adding the shading loss and PV system loss (as instructed by the PV supplier as 10% on average). The system performance ratio on electricity generation was calculated by deducting the performance loss from total. Table 1 shows the system performance ratio across the year in each building group. These values will be used for energy output calculation which is explained in Sect. 3.3.

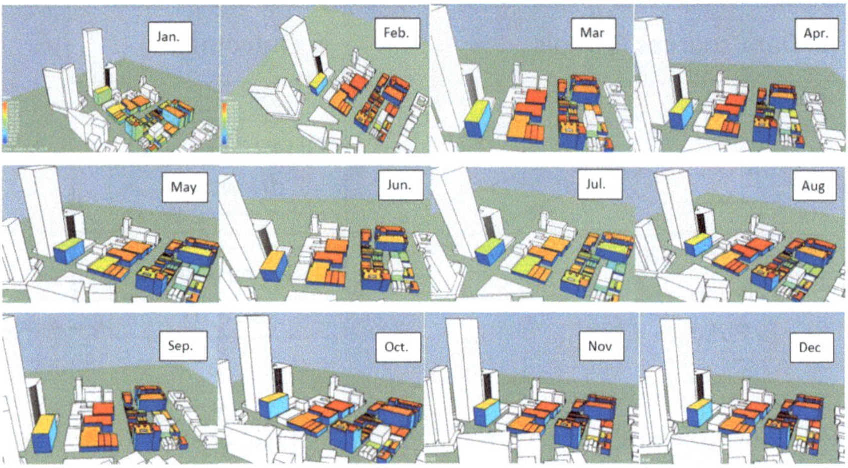

Fig. 2 Monthly solar irradiation value of the case buildings

3.2 *System Design and Capital Costs*

This feasibility study used 250 W polycrystalline solar panels manufactured by Suntech Power.[1] Polycrystalline silicon, or also called polysilicon, uses a polycrystalline form of silicon with a high purity as a raw material [2]. Simpler manufacturing process makes it cheaper to produce polycrystalline solar cells than monocrystalline ones. Polycrystalline cells present a lower cost per watt and have better temperature de-rating coefficient which means they can produce more power in hotter weather [21]. Table 2 shows the technical specifications of the PV systems in this study. The two (2) PV design scenarios for the case study buildings are flat (horizontal) and 30° tilt facing the north (Fig. 3). The flat design has more panels than the tilted design. This is because during the installation of PV systems, flat design has only 0.5 m space between each row, while the tilted design needs to consider about 1 m spaces between each row; otherwise, the panels in front may have shadows over the panels behind. However, the tilted design will get more solar irradiance compared to the flat design, which means that each tilted panel would generate more energy than the flat ones. Therefore, it is meaningful to compare the two design options on the life cycle cost–benefit. Table 3 summarizes the number and size of PV systems.

Table 4 is a summary of capital cost for each design. It should be noted that the cost includes material and installation fees. Most of flat design has higher capital cost than the tilted design (as shown in the last two columns) because of more PV system numbers. Inverter cost of flat design is also generally higher than the tilted design, but only a small proportion in the total system costs.

[1] Suntech Power (http://www.suntech-power.com/) is a leading solar photovoltaic manufacturer in the world. The selected PV product is one of the popular systems applied in Australia.

Table 2 Features and technical specifications of the PV systems

Feature	Description
Solar cell	Polycrystalline silicon 156 × 156 mm (6 in.)
No. of cells	60 (6 × 10)
Dimensions	1640 × 992 × 35 mm (64.6 × 39.1 × 1.4 in.)
Weight	18.2 kgs (40.1 lbs.)
Front glass	3.2 mm (0.13 in.) tempered glass
Frame	Anodized aluminum alloy
Junction box	IP67-rated (3 bypass diodes)
Output cables	TUV (2Pfg1169:2007), UL 4703, UL44
	4.0 mm^2 (0.006 in.^{2}), symmetrical lengths (−) 1000 mm (39.4 in.) and (+) 1000 mm (39.4 in.)
Connectors	MC4 connectors
Maximum power at STC (Pmax)	250 W
Optimum operating voltage (Vmp)	30.7 V
Optimum operating current (Imp)	8.15 A
Open circuit voltage (Voc)	37.4 V
Short circuit current (Isc)	8.63 A
Module efficiency	15.4%
Operating module temperature	−40 to +85 °C
Maximum power at STC (Pmax)	250

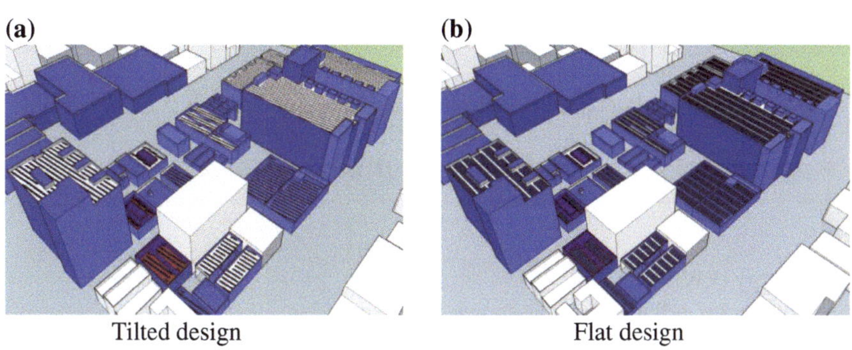

Fig. 3 Design scenarios **a** tilted design **b** flat design

Table 3 Number and size of the PV systems

Group	Flat		30 degree	
	Number	Size (m^2)	Number	Size (m^2)
Building group 1	377	610.7	243	393.7
Building group 2	483	782.5	282	456.8
Building group 3	1637	2652	1051	1702.6
Building group 4	204	330.5	139	225.2
Building group 5	136	220.3	86	139.3

Table 4 Capital cost

Building group	Flat PV system cost (AUD)	Tilted PV system cost (AUD)	Inverter cost of flat design (AUD)	Inverter cost of tilted design (AUD)	Total Flat design cost (AUD)	Total Tilted design cost (AUD)
Building group 1	20,389.6	15,292.2	2400	2000	22,789.6	17,292.2
Building group 2	35,218.4	20,389.6	4800	2400	40,018.4	22,789.6
Building group 3	16,682.4	16,682.4	2000	2000	18,682.4	18,682.4
Building group 4	35,681.8	23,633.4	4800	3000	40,481.8	26,633.4
Building group 5	16,682.4	16,682.4	2000	2000	18,682.4	18,682.4

3.3 Energy Output

The following formula is used to estimate the electricity outputs:

$$E = A \times r \times H \times \mathrm{PR}$$

where

- E Energy (kWh);
- A Total solar cell area (m^2, as shown in Table 3);
- r PV product efficiency (15.4%);
- H Hourly Solar radiation (extracted from PVWatts);
- PR Performance ratio, coefficient for losses (as shown in Table 1).

The results show that flat design generally generates more energy than the tilted panels (Table 5) by 50%. This is mainly due to the larger number of PV systems which can be installed in the flat design layout. However, the cost of the flat PV panel installation (Table 4) is also higher than tilted panels by 40%. A lifecycle

Table 5 Energy output in year 1

Buildings	Flat (KW)	Tilted (KW)
Building group 1	10,5917.71	47,199.12
Building group 2	10,2705.78	59,964.86
Building group 3	52,5735.23	311,648.35
Building group 4	61,139.09	29,453.94
Building group 5	36,433.85	15,932.19

cost–benefit assessment is necessary. With PV system performance attenuation, the net electricity export in each year will decrease. This has also been considered in the cost–benefit assessment of this study. The 25-year energy output is calculated based on year 1 data by using a 0.5% system attenuation ratio (as instructed by the PV supplier).

3.4 *Cost–Benefit Analysis*

Calculation of benefits is based on the actual peak and off-peak electricity tariff rates and the feed-in tariff rates. Two types of benefits are calculated:

- Saved Electricity Tariff: The savings obtained by supplying the electricity produced from the PV system to the buildings as per the demand;
- Gains from feed-into the grid: Extra energy produced from the PV system, after satisfying the building energy demands. This extra energy is sent to the public grid as feed-in energy. The feed-in tariff rate 5 c/kWh in Victoria, for large-scale projects [20], is considered for this study.

A customized spreadsheet tool was used to calculate the "Saved electricity Tariff" and "Gains from feed-in" for every 15 minutes for 25 years as shown in Table 6. The savings of each year were used to calculate the Net Present Value (NPV) of the system and the payback period.

The following equations explain the formula used for the above scenarios:

Table 6 Example excel template used

Scenario	Energy output	Energy consumption	Extra energy required from public grid	Extra energy for feed-in	Base savings by supplying to the buildings	Savings from feed-in
1	0.00	13.180	13.180	0.00	0.00	0.00
2	36.50	13.054	0.00	23.446	2.48	1.172
3	2.976	13.657	10.681	0.00	0.57	0.00

Where Energy Output < Energy Consumption,
Total Savings = Saved Electricity Tariff
Saved Electricity Tariff = Energy Output * applicable electricity tariff rate
Where Energy Output > Energy Consumption,
Total savings = Saved Electricity Tariff + Gains from feed-in
Saved Electricity Tariff = Energy Consumption * applicable electricity tariff rate
Gains from feed-in = Extra Energy available * feed-in tariff rate

The Net Present Value (NPV) is used to present value of a series of future cash flows in the alternative designs:

$$\mathrm{NPV} = -C_0 - \sum_{n=10}^{T} \frac{M_1}{(1+r)^n} + \sum_{n=1}^{T} \frac{C_1}{(1+r)^n}$$

where C_0 = net construction costs; $M_{10\&20}$ = maintenance costs; $C_{1\text{–}25}$ = electricity savings; r = the discount rate.

- The net construction costs include product, transportation, and installation expenses (as shown in Table 5).
- The maintenance costs are mainly related to the replacement of centralized inverters every ten years.

To properly calculate the current value of the PV designs at initiation, this study adopts the Discount Cash Flow model. The discount rate is in accordance with the required rate of return of the project or at least the financing cost of the project. Since the case buildings would be invested fully by the project client, it is assumed that there is no risk premium in the discount rate, and the client finances the project entirely with the proceedings of municipal bond around that time. Table 7 shows the cost–benefit of each building group during the 25 years, and Table 8 shows the NPV results. The results are discussed in the next section.

3.5 *Key Observations from the Cost–Benefit Assessment Results*

Table 8 shows the NPV of applying PV systems in the selected high-density regions of Melbourne. The results indicated the positive directions of distributed PV adoption. In the last decade, photovoltaic technologies have experienced unprecedented cost reductions among electricity-conversion technologies. The large-scale uptake of solar panels in buildings has reached an era which relies more on optimized design instead of financial support from the government. This gives confidence to building clients on the usage of renewable energy in overshadowing areas.

Table 8 also shows that the flat design is better than the tilted design from client's economic perspective. This is because the flat design can use more solar panels on

Table 7 Cost–benefit of the PV systems during 25 years

Cost–benefit—flat (AUD)						Cost–benefit—tilted (AUD)					
Year	Building group 1	Building group 2	Building group 3	Building group 4	Building group 5	Year	Building group 1	Building group 2	Building group 3	Building group 4	Building group 5
1	17,744	17,141	88,488	10,185	5462	1	7909	10,073	52,375	4933	2577
2	17,448	16,863	87,005	10,020	5363	2	7775	9904	51,503	4850	25,33
3	17,965	17,357	89,600	10,314	5481	3	8006	1019,7	53038	4994	2602
4	18,498	17,867	92,273	10,617	5605	4	8243	10,499	54,620	5141	2673
5	19,046	18,392	95,025	10,930	5732	5	8487	10,810	56,249	5294	2747
6	19,611	18,933	97,859	11,252	5865	6	8739	11130	57,926	5451	2823
7	20,193	19,491	10,0778	11,584	6002	7	8998	11,460	59,654	5612	2901
8	20,792	20,066	103,783	119,26	6144	8	9264	11,800	61,434	5779	2982
9	21,409	20,658	106,879	12,278	6291	9	9539	12,150	63,266	5950	3065
10	22,044	21,269	110,066	12,642	6443	10	9822	12,511	65,154	6127	3151
11	22,699	21,897	113,349	13016	6601	11	10,113	12,882	67,098	6308	3239
12	23,373	22,545	116,730	13,401	6765	12	10,414	13,265	69,100	6496	3331
13	24,067	23,213	120,211	13,798	6934	13	10,723	13,658	71,162	6688	3425
14	24,782	23,901	123,796	14,207	7109	14	11,041	14,064	73,285	6887	3522
15	25,519	24,610	127,489	14,629	7291	15	11,369	14,482	75,472	7091	3622
16	26,277	25,340	131,291	15,063	7479	16	11,706	14,912	77,724	7302	3726
17	27,058	26,092	135,207	15,511	7673	17	12,054	15,355	80043	7519	3833
18	27,862	26,867	13,9239	15,972	7874	18	12,412	15,811	82,432	7742	3943
19	28,690	27,666	143,392	16,446	8082	19	12,781	16,281	84,891	7972	4056
20	29,543	28,489	147,669	16,936	8297	20	13,160	16,765	87,424	8209	4173
21	30,421	29,337	152,073	17,440	8519	21	13,551	17,263	90,032	8453	4294
22	31,326	30,210	156,608	17,959	8750	22	13,954	17,777	92,718	8704	4418
23	32,258	31,110	161,279	18,494	8988	23	14,369	18,305	95,484	8963	4547
24	33,217	32,037	166,089	19,045	9234	24	14,796	18,849	98,332	9229	4679
25	34,205	32,992	171,043	19,612	9488	25	15,236	19,410	101,265	9504	4815
Summary	616,046	594,343	307,7220	353,276	177,469	Summary	274,460	349,614	182,1679	171,197	87675

Table 8 Net Present Value

Discount rate 3%				
Group	NPV—flat (AUD)	Payback year—flat (Year)	NPV—tilted (AUD)	Payback year—tilted (Year)
Building group 1	192,183	13	418,12	19
Building group 2	115,779.4	17	69,685.1	17
Building group 3	1101,952	11	605,400.3	12
Building group 4	115,640.6	13	32,216.1	17
Building group 5	38,589.3	16	7664.7	22

the same size of roof spaces compared to tilted design. This finding is different from the rule of thumb used in PV designs that the tilted degree is consistent with the local altitude. Through just one case, we cannot claim that the flat design should be promoted; however, it shows the importance of comparing the alternative designs on tilted degrees to maximize the economic outcomes. The third building group has the shortest payback period (i.e., 11 years) in the flat design (25 years of total lifespan is assumed). This is due to the lower shading loss and larger designed PV size, which indicate the significance to evaluate the impact of surroundings and maximize design areas on buildings. With the rapid urbanization in Melbourne CBD, urban planning and approval process should incorporate comprehensive modeling and value assessment of the renewable energy systems to inform decision making.

4 Conclusions

Current approaches for the reduction of carbon emissions in buildings are often predicated on the integration of bespoke renewable technologies into building projects. With access to abundant solar resources, solar energy is an attractive option in Australia among various renewable energy sources. We conducted a detailed value assessment to optimize the cost of applying PV systems in a high-density city area of Melbourne. The NPV results evidence the feasibility to apply roof-mounted polycrystalline PV products in the case study buildings. In general, the payback period of the PV systems in this study is around 15 years. The flat design generally generates more energy than the tilted panels by 50%. This is mainly due to the larger number of PV systems which can be installed in the flat design layout. However, the cost of the flat PV panel installation is also higher by 40% compared to tilted panels. Through the NPV assessment, this study also shows that the flat design is better than the tilted design from client's economic perspective. This research supports investors' decisions by understanding the financial values of prefabricated PV systems in high-density regions and provides suggestions to building professionals on value-for-money design. It also sheds light on the opportunities on the uptake and diffusion of PV and bespoke low-carbon technologies in general.

References

1. APVI (2015) Summary of Australian PV industry survey. http://apvi.org.au/. Accessed Dec 2017
2. Bagher AM, Abadi Vahid MM, Mohsen M (2015) Types of solar cells and application. Am J Opt Photon 3(5):94–113
3. Bakirci K (2012) General models for optimum tilt angles of solar panels: Turkey case study. Renew Sustain Energy Rev 16(8):6149–6159
4. Benghanem M (2011) Optimization of tilt angle for solar panel: case study for Madinah, Saudi Arabia. Appl Energy 88(4):1427–1433
5. Buckman G, Diesendorf M (2010) Design limitations in Australian renewable electricity policies. Energy Policy 38:3365–3376
6. City of Melbourne (2015) Daily population estimates and forecasts. https://www.melbourne.vic.gov.au/SiteCollectionDocuments/daily-population-estimates-and-forecasts-report-2015.pdf. Accessed 20 Dec 2017
7. Climate Commission Australia (2013). https://climatecommission.angrygoats.net/report/solar-energy/. Accessed 29 July 2017
8. Department of Environment and Energy (2017). http://www.environment.gov.au/climate-change/government/renewable-energy-target-scheme. Accessed Dec 2017
9. Department of industry innovation and science (2017). https://www.industry.gov.au/resource/Mining/AustralianMineralCommodities/Pages/Coal.aspx. Accessed Dec 2017
10. Energy Networks Australia, CSIRO (2016) Electricity Network Transformation Roadmap: Key concepts report 2017–27
11. Gregg A, Parker T, Swenson R (2005) A "real world" examination of PV system design and performance. Photovoltaic Specialists Conference, 2005. Conference Record of the Thirty-first IEEE, 2005, 1587–1592
12. Hussein HMS, Ahmad GE, El-Ghetany HH (2004) Performance evaluation of photovoltaic modules at different tilt angles and orientations. Energy Convers Manage 45(15–16):2441–2452, ISSN 0196-8904. https://doi.org/10.1016/j.enconman.2003.11.013
13. IES-VE Suncast. https://www.iesve.com/software/ve-for-engineers/module/SunCast/457. Assessed 20 Jan 2017
14. IPCC (2014) Fifth Assessment Report. Climate Change 2014: Working Group III: Mitigation of Climate Change. Retrieved from Valencia, Spain
15. Jantsch M, Stoll M, Roth W, Kaiser R, Schmidt H, Schmid J (1991) The effect of tilt angle and voltage conditions on PV system performance an experimental investigation. In: Luque A, Sala G, Palz W, Dos Santos G, Helm P (eds) Tenth E.C. Photovoltaic Solar Energy Conference. Springer, Dordrecht
16. NREL (2005) National solar radiation data base daily statistics files. National Renewable Energy Laboratory. http://rredc.nrel.gov/solar/old_data/nsrdb/dsf/. Accessed 30 June 2015
17. PVWatt Calculator (2017) http://pvwatts.nrel.gov/. Assessed 20 Jan 2017
18. PVsyst (2017) http://www.pvsyst.com/en/. Assessed 20 Jan 2017
19. Rowlands K, Beausoleil-Morrison (2011) Optimal solar-PV tilt angle and azimuth: An Ontario (Canada) case-study. Energy Policy 39(3):1397–1409
20. Victoria State Government (2017) https://www.energy.vic.gov.au/renewable-energy/victorian-feed-in-tariff. Accessed 25 Dec 2017
21. WBBCC (2010) The differences between crystalline and thin film, mono and poly etc., Solar Technology explained, WBBCC 2010. https://wbbcc.files.wordpress.com/2010/08/solar-technology-explained.pdf. Accessed 29 July 2017

22. Yadav Amit Kumar, Chandel SS (2013) Tilt angle optimization to maximize incident solar radiation: a review. Renew Sustain Energy Rev 23:503–513

Dr. Rebecca Yang B.Sc., M.Sc., Ph.D. is a senior lecturer and scholar of building, construction and social practice. She undertakes pure and applied research that can provide innovative solutions to the industry by integrating theories with cutting-edge technologies.

Mr. Andrew Carre B.E.(Hons.), B.Ec., M.E. is a lecturer and an energy engineer with expertise in life cycle assessment. He has considerable experience in research and client management.

Are We Living with Our Heads in the Clouds? Perceptions of Liveability in the Melbourne High-Rise Apartment Market

Sarah Holdsworth, David Kenny, Jeremy Cooke and Shaun Matfin

Abstract Housing in the Melbourne metropolitan area is in the midst of a push towards intensification through increased densification of high-rise apartment dwellings. This reflects similar international trends in housing provision, a consequence of increasing global populations and the need to intensify land use in the quest for more sustainable urban areas. However, the Melbourne housing market is inexperienced in the planning, design, delivery and habitation of high-rise development. Evolving planning legislation, which draws on existing international high-rise planning policy, recognises that current developments entering the market are lagging behind international standards in relation to the degree of liveability these buildings afford residents. This chapter examines the characteristics of liveability and design in the context of high-rise residential developments which include consideration of building amenity, apartment amenity and external amenity. It then presents the findings of 13 semi-structured interviews with key stakeholders involved in the design and construction of high-rise apartments in Melbourne's CBD. The interviews explore perceptions of liveability as they inform and consequently manifest in current projects. The findings identified that liveability is a subjective term encompassing a variety of characteristics which different stakeholder groups emphasised differently based on their disciplinary background. The findings are important as there exists a limited understanding of how the industry conceptualises high-rise developments and in turn makes design and development decisions in the context of liveability. Further, it was recognised that all participants wanted to improve the liveability of their development and were prepared to collaborate across discipline to achieve such outcomes. This goal will not be achieved if interdisciplinary understandings are not identified, shared and built into the process.

S. Holdsworth (✉)
Sustainable Building Innovation Laboratory, School of Property Construction and Project Management, RMIT University, Melbourne, VIC, Australia
e-mail: sarah.holdsworth@rmit.edu.au

D. Kenny · J. Cooke · S. Matfin
School of Property Construction and Project Management, RMIT University, Melbourne, VIC, Australia

P. Rajagopalan et al. (eds.), *Energy Performance in the Australian Built Environment*, Green Energy and Technology, https://doi.org/10.1007/978-981-10-7880-4_12

1 Introduction

Global urbanisation has the potential to usher in a new era of well-being, resource efficiency and economic growth. Achieving these outcomes will require developers and planners to address conflicting and divergent problems in the areas of protecting the natural environment while promoting the economically growing city and advocating social justice [2]. Global approaches to environmental limits and population growth in urban areas have resulted in a push towards increased urban consolidation through densification and intensification and have manifested in the development of high-rise apartments. While this addresses many environmental sustainability issues, e.g. decreasing urban sprawl and increasing the energy and water efficiency of housing through the provision of green buildings and compact cities, it is reshaping how people live and experience life. It is important to recognise that sustainability issues are simply limited not to environmental impacts, but also to the development of social and human capital. Sustainable cities must consider how people experience the built environment through the provision of appropriate amenity. Sustainability issues will not simply be solved by land intensification; poor design outcomes may result in increased negative impacts on the environment and result in poor social cohesion and dislocation.

In Australia, approximately 90% of the population live in urban areas, making it one of the most urbanised nations in the world [26, 27]. Of all Australian capital cities, Melbourne will see the largest population growth with the population predicted to grow by 3.9 million from 2011 to 2051 [1, 12]. This growth will place a strain on the existing residential building stock, associated amenity and will require an additional 1.6 million dwellings [1, 12]. Melbourne's current strategic development plan, Plan Melbourne 2017–2050, seeks to manage population growth through increased densification in the central business district (CBD) (and other strategic city locations) and improved liveability [11]. However, a number of factors including an understanding of the term "liveability" in a high-rise context, associated delivery methods and financial incentives make this a difficult vision to achieve [4, 6].

Currently, loosely regulated development coupled with a focus on investment returns and growing demand for accommodation, facilitating residential development opportunity, has resulted in a decline in the standard of housing within high-rise development, increased pressure on public infrastructure and the associated amenity and liveability [5, 7, 9]. These outcomes were illustrated in a survey undertaken by the City of Melbourne [9] which analysed 3500 apartments using the UK's Commission for Architecture and the Built Environment (CABE) design criteria to measure the sustainability of residential development. The report found only 16% of apartments scored a "good" outcome while 33% scored a "poor" outcome. Further, it can be argued that these high-density apartment blocks have become mechanisms for short-term profit generation at the expense of human and social capital. While increasing Melbourne's housing stock without contributing to the existing problem of urban sprawl can make this type of development appear attractive, the effects on their residents must be taken into consideration.

The increased dominance of high-rise habitation will not only be felt by the current residents but for future generations, of which high-rise apartment living will become a way of life. With the high development costs and long lifecycles of these buildings, it is important that these developments are constructed to cater for the needs of a diversity of residents along with the wider society now and into the future. Liveable communities are those that provide individuals with choices, opportunities that are affordable, accessible, socially inclusive, safe and allow them to fulfil their potential [23]. Good design principles inform the liveability of communities as the paradigm refers to the design of dwellings which are flexible and adaptable over time, designing liveable spaces and urban environments which contribute to the social and human capital of those who inhabit them [10, 13]. How we access and experience the built environment will inform the liveability and quality of our lives.

This chapter builds upon previous chapters (see Chapters "Low-Energy Housing as a Means of Improved Social Housing: Benefits, Challenges and Opportunities" and "Indoor Environmental Quality of Preparatory to Year 12 (P-12) Educational Facilities in Australia: Challenges and Prospects") and explores the concept of liveability within Melbourne's high-rise metropolitan area, specifically the perception of liveability in the context of high-rise apartments, as it is understood by key industry stakeholders in Melbourne's high-rise apartment market. Liveability relates not only to improved indoor environments, i.e. quality and efficiency, but also to the design as it informs lived outcomes within the apartments and surrounds in terms of amenity and its proximity afforded to residents. This chapter begins with the examination of the variables that contribute to liveability and presents a conceptual liveability framework in the context of high-rise development. The variables are then explored through a series of semi-structured interviews with key stakeholders involved with the development process of Melbourne's apartments, and the results are discussed in the context of the conceptual framework.

2 Density, Crowdedness, Intensity and Liveability

Density, crowdedness, intensity and liveability are all related terms, but how they relate and manifest in the urban environment is subjective. Density, in an urban content, is defined by the number of units in a given area [15]. It does not inform how an urban place feels or functions and this influences the resultant environmental and social impacts (the sustainability of the built form). High density in cities with little or no precedent for living in high-density environments has become a byword for cramped, noisy accommodation, where one's quality of life suffers accordingly (despite the potential for environmental improvement/efficiency through a decreased ecological footprint) [18]. Forsythe [15, p. 2] surmises that "increased density is feared by those who imagine ugly buildings, overshadowed open space, parking problems, and irresponsible residents". It can be assumed that the success of high-density developments is informed by not only the design of the dwellings but also the urban context and associated amenity in which they are located (such as schools, shops, medical and recreational facilities).

Recently, building density levels in the Melbourne CBD have seen ratios of 30:01 and as high as 55:01 [17]. However, when referring to building density caution must be taken as a building may be densely populated with small dwellings that are sparsely inhabited which would result in a low degree of crowdedness, while still being perceived as a building of high density. Crowdedness is the correlation between the liveable space and building density [32]. Yeh [31] surmised that with the correct planning and development mechanisms in place to inform design and construction, high-rise developments have the potential to offer "good density", reduce levels of crowdedness and improve the urban living environment [32]. Similarly, urban intensity is related to density; Landcom [20] articulates that intensity relates to how an individual responds to an urban environment. Crowdedness and intensity are often referred to in the literature as the key issues surrounding perceptions of high density and are more reliable measures of liveability than density as they reflect how a residential development feels or is experienced [20, 32].

Liveability, like crowdedness and intensity, is a subjective term and difficult to define; Buys et al. [8] argues that "*what constitutes a liveable place is very complex, very personal and therefore difficult to articulate*". Liveability in an urban context can relate to both the liveability of the individual dwellings and the liveability of the urban areas at large. Important to the liveability of dwellings are principles of good design which are essential to the development of urban form that make a positive contribution to the health and well-being of both neighbourhood, homes and their inhabitants [9]. Good design principles in the construction of dwellings refer to a range of factors such as layout, size and configuration within an apartment, allowing for future adaptability, and ensuring maximum liveability for the occupier, rather than simply meeting minimum building or planning regulations. Additionally, good design includes the relationship between the dwelling, the environment and associated amenity. This is reflected in the desire for compact cities embedded in much of the global planning literature and associated policies. Amenity can be divided into two categories: external and internal. Internal amenity refers to how the residential space is designed and the degree to which that space is perceived as liveable [17]. External amenity relates to the urban built environment outside the living space such as open green spaces and how they provide for the satisfaction of residents [31]. Compact cities are predicated on amenity that decreases the need for travel and therefore improving the environmental impacts of population growth as well as quality of life [11, 17].

3 Melbourne's CBD, High Rise, Density Development and Liveability

In an effort to curb Melbourne's growing urban sprawl and housing shortfall, the Victorian state government's strategic planning policy has focussed on increasing the number of dwellings in the CBD [5, 6]. However, loosening of planning controls combined with neoliberal market-driven economic policy has allowed developers to determine housing type and timing within the market [5, 22, 24]. This has resulted in the domination of a single model: small one and two bedroom apartments in high-rise blocks. These newly completed developments have been negatively critiqued for a lack of liveability for the occupier [9, 24]. Further, it has been argued that the poor quality of design in Melbourne's existing high-density apartment development is a consequence of developers simply complying with Building Code of Australia standards rather than having to conform to standards enforcing good design principles [9].

Current global guidelines emerging for higher density, high-rise buildings focus on areas such as urban context, building envelope and green space inclusion when attempting to regulate for increased standards of liveability [17]. However, what has been shown to be far more important is whether the building delivers good living outcomes for the residents and whether it has a positive impact on the broader community [17]. Evidence from countries that have successfully regulated for good apartment design and increased levels on liveability, such as the UK and Hong Kong, show a concerted focus on establishing appropriate density controls, density reduction incentives for developers to deliver public spaces, regulated tower separation rules, building amenity and the establishment of minimum apartment design standards [17].

Comparing Melbourne's current standards to other large cities such as London and Sydney, it has been identified that Melbourne has the least rigorous policy guidelines on housing quality, specifically in the area of measurable outcomes including minimum apartment sizes, requirements for the orientation of apartments and minimum internal amenity standards [9, 17]. However, in an effort to promote liveability in new apartment developments, Clause 58 Better Design Guidelines was introduced in 2017 into the Victorian Planning Scheme. These guidelines aim to improve the overall functionality of new apartments and include a focus on apartment layout, minimum ceiling heights, adequate provision of storage and access to natural daylight and ventilation. There is no mandatory guide for size, space between buildings or internal building amenity. It is proposed that Clause 22.01 Urban Design within the Capital City Zone and Clause 22.02 Sunlight to Public in the City of Melbourne Planning Scheme will provide direction on building separation, plot ratio and overshadowing [11].

4 How to Measure Liveability?

For the purpose of understanding and measuring liveability from a residential perspective within high-density development, this research draws primarily on the work conducted in Brisbane by [8], Singapore by Wong [30] and Dublin by Howley et al. [18]. These studies collected primary data from residents of high-density developments critiquing various aspects of high-density living. The positive and negative contributors of liveability in high-rise living identified by the authors can be found in Table 1.

Table 1 Positive and negative contributors to liveability in high-rise apartments

Source	Positive factors identified	Negative factors identified
Buys et al. [8]	Apartment design: private open space (i.e. balcony) passive cooling and favourable solar aspects Diversity of local amenities Lack of social engagement, residents identifying anonymity and a certain isolation within the building's community as an appealing aspect of high-density living	Noise, especially from other residents, and to a lesser extent from external sources such as traffic, as the biggest nuisance of high-density living Lack of facilities and a reluctance to use the buildings common amenities Lack of public transport services
Howley et al. [18]	The study recognises a significant level of dissatisfaction of residents living in new high-density areas	Quality of the apartment Noise, including traffic noise, noise from anti-social behaviour and construction work Pollution Lack of amenity, specifically shopping facilitates a lack of facilities for children and a desire for more green space Lack of social engagement within the building
Wong [30]	Better view Fresher air More windy Quieter environment High-rise living as a lifestyle Better quality of housing High level of surrounding local amenity	Safety of the building structure Ease of escaping in emergency Longer waiting time for lift Lack of community interaction Insufficient supporting facilities Greater danger of high-rise littering Personal fear for height Higher pricing

The work of Liu [16], Williamson [21], Gifford [29] identified similar trends recognising residential satisfaction or perception of liveability directly relates to apartment size, spaciousness, layout, building aesthetics, quality of construction and the surrounding neighbourhood. Forrest and Kearns [14], Myers [25], Whyte [28] concluded that residential satisfaction not is just limited to the internal living space but may also include the external space and surrounding amenity. These findings further highlight the significance of the individual dwelling, the building itself and the surrounding amenity as being pivotal to the creation of what residents perceive to be a liveable space.

At present, there is no single definition that constitutes a liveable space and accounts for quality of life within that space [8]. However, to provide a starting point for this research in the Melbourne context, we have developed a conceptual liveability framework, from the literature, to explore how decisions are made about the design of a liveable apartment. This framework is presented in Fig. 1. It has been developed from the literature and structured on three components:

1. Internal amenity (the dwelling itself)
2. Building amenity (the apartment building)
3. External amenity (the immediate environs of the building).

5 Research Method

Based on the findings from the literature review, 13 semi-structured interviews were undertaken with key industry figures involved in the development process; developers, planner/urban designers, architects and builders were undertaken, a method well suited to exploring the perceptions and opinions of the respondents regarding complex issues [19]. A purposive sampling strategy, drawing on known contacts, was applied to the research. Each of the individuals interviewed held senior roles within their respective industries, with over 15 years of experience; see Table 2. This allowed for the exploration of the complex issues surrounding liveability in the Melbourne high-rise apartment market. The interview questions were structured to explore the following topics:

- Perception/definition/characteristics of liveability
- Relationship between good design and liveability
- How decision is made about the design of high-rise apartment buildings.

Interviews were digitally recorded transcribed, and data was subject to thematic analysis.

Table 2 Interview participants

Stakeholder group	Developers	Architects	Planners	Builders
Number of participants	4	4	2	3

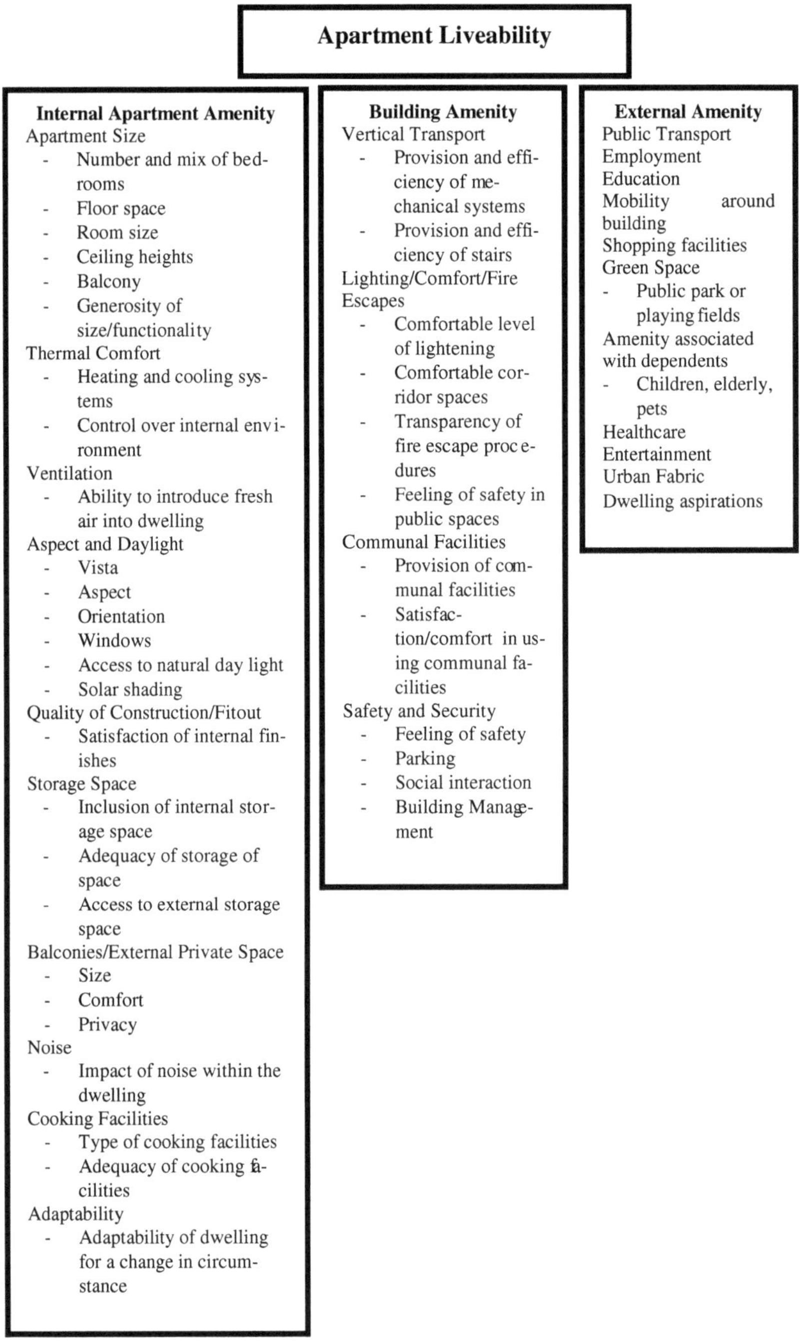

Fig. 1 Conceptual framework of liveability in high-rise, high-density living

The small sample and the aim of research were not to try for a comprehensive study, but with the limited resources available, to undertake a scoping study to gain an understanding of the issues, that could provide the base of a more extensive investigation. Additionally, the scope of this research only included those involved in the design and construction of high-rise apartments. Further research to investigate how residents experience such developments provides for a further avenue of investigation.

6 Results

6.1 Perception of Liveability

In the context of high-rise apartment living, participants were asked to define liveability. Interestingly, none of the participants was able to articulate a clear definition of liveability. However, all participants recognised the importance of liveability in relation to the impact it had on day-to-day lives of the "end user":

> Liveability to "them" is how they are able to live their life, not just where they have to sleep (P 7, architect).
>
> it may well be about, for some, accessing services. It may well be, to others, general space they have within their dwelling, or, it may well be access to a lifestyle (P10, developer).

When participants reflected on aspects of "liveability" they did so, not from their own "lived" experience, but from their imagined experience of those living in apartments. Participants were asked about the key variables that contributed to the liveability of high-rise apartments, their responses correlated to their disciplinary experience. Developers broadly identified aspects such as "location" or "lifestyle" (P4, Developer) as contributing factors to a good life. As such, much of their discussion focussed on the location and internal amenity of the high-rise building itself (e.g. pool, roof top garden). Architect responses focussed on the apartment design and identified factors such as "sunlight" and good "ventilation" (P5, Architect) as key contributors to high levels of liveability. However, similar to developers, architects also referred to the fact that a building's location was an important factor when they considered liveability. The components of liveability when examined from the perception of planners/urban designers focussed mostly on the apartment design: "size, quality and type of construction" (P4, planner), and the relationship between the building and the surrounding urban form "environment or neighbourhood" (P8, planner). Builders placed the greatest importance on both internal and external amenities such as "natural light", "privacy" (P3, builder), "adequate space" (P12, builder) and the "sense of community" afforded by the building itself (P3, P11 and P12).

As in the literature, a clear definition was difficult for participants to articulate. Further, it was apparent that the understanding of liveability is inherently related to an individual's perspective as informed by their background.

Table 3 Characteristics of good design in high-rise apartments

	Developers (n = 4)	Architects (n = 4)	Planners/urban designers (n = 2)	Builders (n = 3)	Total
Characteristics of liveability	Frequency of citations				
Functionality of space within the apartment	10	12	4	7	31
Apartment design characteristics natural light Ventilation Noise insulation, Storage	Total 21 10 4 5 2	Total 17 6 5 4 2	Total 8 4 2 1 1	Total 7 2 0 5 0	74
External amenity (Surrounding community and services)	7	4	5	13	29
Building amenity	21	8	2	12	43
Affordability/price in the consideration of design	5	9	6	16	36

6.2 Internal Apartment Amenity and Liveability

A number of key building characteristics were identified by participants in relation to liveability. The frequencies of characteristics are summarised in Table 3.

From the responses provided, apartment design was the most identified characteristics of liveability articulated by all stakeholders. In order, functionality of space, building amenity and external amenity were the most frequently noted characteristics. The impact of cost of land, finishes and construction were most frequently cited as impacting on design decisions. With affordable housing correlating to a decrease in apartment size, one architect commented "*I think the smaller space is about being more affordable*" (P5). The data revealed that all participants cited good apartment design as the most significant factor in achieving a good lived experience, as stated "*good design will make good liveability. Good design is not just the physical appearance, it's the design of the space, the design of the apartment*" *(P6, architect).* Participants defined "good apartment design" as characteristics of internal amenity including: access to natural light, ventilation for fresh air, quality, noise insulation and sufficient storage. This is illustrated by the following responses:

> quality of an apartment is around having, within that space, good natural light, good storage (P10, developer).
>
> I'd rather have a dwelling that had been thought about in terms of the quality of it, in terms of the light, and the ventilation and the actual usefulness of space (P7, architect).

Participants were asked whether good design could negate decreased apartment size, and responses revealed the largest polarisation between the different professions.

Developers held the opinion that good design was vital to liveable outcomes and could compensate a lack of space, as illustrated by the following response "it doesn't matter how small we go as long as it's functional" (P2, Developer). The other stakeholder groups, planners, architects and builders did not share this opinion, as evident by the following quotes:

> "absolutely not", "size is size…you need space." (P4, planner).
>
> you can have all the wonderful finishes you want, but if it's small and pokey…it's not liveable (P5, architect).

Two builders described small apartments as "*hotel rooms*" and somewhere you could "*sleep, but not live*". These comments reflect a concern held by all professions, other than developers, that a loss of size often directly resulted in a loss of functionality. Architects were most concerned with functionality and emphasised the need for better designs to offset reduced apartment sizes "*smaller compact living is fine, if it's well designed and the functions work*" (P7, architect). One planner/urban designer noted that size of apartments related to "*having enough space for the basic functions*" (P8 planner/urban designer).

6.3 *Building Amenity and Liveability*

Amenity, and associated lived experience within a building, was identified as an important factor associated with liveability. Of the stakeholders, developers most frequently identified building amenity as critical for the creation of a lifestyle for residents citing the range of amenity items offered within their buildings including pools, gyms, club rooms, kitchens, Mah-jong rooms, billiard rooms, libraries, cinemas and golf driving ranges. These items were also recognised as a selling point as their inclusion was not regulated. While architects shared the belief of the importance of such amenity, they also recognised the importance of spaces that were flexible in use to suit the needs of a variety of residents. One architect commented it is "*the space that is the value not the facility they [developer] put in there*" (P7). However, increased building amenity was recognised to result in excessive body corporate fees "*communal facilities can be very expensive to maintain*" (P8, Planner/urban designer).

6.4 *External Amenity*

External amenity was perceived by participants as being a necessary aspect of any definition of liveability. External amenity was identified as the urban built environment outside the living space, such as access to public transport and services. Given these various elements of external amenity are subjective in how they provide for people's enjoyment and satisfaction, the responses from participants varied. One of the responses included, "*Liveability for an apartment dweller is around a lifestyle choice that says, 'we require access to services, we have a lifestyle where we don't spend our entire time in our back yard, we don't have one!'*" (P10, developer). As one of the participants stated "*people look for what they consider to be a good location*". The focus on external amenity by developers was reflected in the participant's responses, "*we are trying to choose sites in areas that have existing amenity or will have existing amenity*" (P2, developer). Of the four groups, builders referred to external amenity with the greatest frequency. This group recognised the important relationship between the building and its neighbouring properties or public spaces in the creation of community.

6.5 *Influences Over Decisions that Impact on Design and Liveability*

The responses of the participants indicated a range of factors that impact on the resultant urban form, including government planning and building policy and regulations, finance requirements, quality of construction and contracts. However, the most frequently noted factors are presented in Table 4.

Not surprisingly the impact of construction costs, land and material as they affect the final dwelling price and planning regulations were most frequently cited as influencing design decision. Following on was banking and access to finance, participants articulated that banking institutions in Melbourne are effectively regulating apartments size: one participant commented *"Banks won't lend people money to buy them if they're not 40 square metres, so therefore developers will make us do that as a minimum" and "the banks seem to be the best regulator of all of this" (P5, architect).* In addition, another participant's response noted financial institutions requirements for presales in order to secure funding: "*most developers are looking to debt finance the development, and in order to get the debt financed from the lending institution, they*

Table 4 Factors influencing the design of high-rise apartment buildings

Factor	Number of respondents (N = 13)
Sale price	10
Bank lending	6
Other developers/similar developments	1
Planning regulations	7
Building regulations	5
Purchaser needs	4
Client	3
Sales agents	2

need to make a certain number of presales" (P7, architect). Presales are apartments sold off the plan and often the bank requires 70–90% sold before releasing funds. It was noted that this inhibits innovation in design, as sales agents require designs that are "guaranteed to sell" as illustrated by one participant about design outcomes "*I don't think they care, I think it's about getting and achieving those presales as quickly as possible, while they're holding onto the land*" (P7, architect).

The nature of the developer's contract with the appointed builder was also recognised as impacting on liveability, in particular the "design and construct" contract. Specifically, the issue is of value management where the design responsibility of the developer's consultant architect is novated to the appointed builder who then assumes the responsibility for the remaining design work of a given project. To ensure the builder maintains, or increases, a desired profit margin as part of their tender process, "value management" practices are undertaken, which can range from altering aspects of the building's engineering to improve constructability, to the substitution of nominated products for cheaper, often inferior quality alternatives, thus impacting on the overall liveability of a building. Comments from all stakeholders unanimously agreed that this approach often led to a reduction in the finished quality of a building. Participant 12, a builder, describes this as process as "*reducing the quality to the minimum that can be achieved*". The effects of this often led to the "*dilution of some of the conceptualised... good design components*" (P8, planner).

The role of a building code was recognised as critical to ensure buildings are constructed to a high standard (see also Chapter "Urban Climate in the Transformation of Australian Cities"). Participants were asked whether they believed current building standards were adequate to provide a good standard of building and therefore liveability, with mixed results, one participant stating yes, seven sufficient and five replying they did not think they were sufficient. One perspective, from a builder was that the current regulations were "*sufficient as a minimum*" (P12, Builder) and that there were "*onerous restrictions*" (P11, Builder). However, planners and architects noted that "*just because that'll do, doesn't mean it's good*" (P5, Architect) and that they believed the "*standard of construction is just awful" (P4,).*It was further noted that the "*government hasn't got enough money to have enough people out there to inspect*" (P13, builder) highlighting examples of companies that were retrofitting building just 8 years after construction. These contrasting views pose an issue for the industry as increased regulation can be costly and limit creativity and design; however, the trouble is if you don't regulate, "*at the edge you get the "dodgy" and the cowboys"* (P1, Developer).

7 Discussion

The research findings concur with that of the literature: liveability is subjective and further relative to place, person, values and experience [3, 25]. Participants were unable to clearly articulate definitions, commenting how difficult it was to comprehensively define liveability in the context of high-rise apartment living. Participants were only able to offer characteristics that they believed contributed positively to the "lived" experience. Consequently, current designs are developed from data collected through real-estate and sales agents that are therefore undertaken retrospectively.

The conceptual model of liveability developed from the literature review identified three key components of liveability in high-rise apartments. The thematic analysis of the interview participants' responses concurred with the findings of the literature review. However, the data analysis revealed that not all of the characteristics of liveability are valued equally by different professions. Consequently, when different professions discuss and make design decisions, they do so from different theoretical bases. This is important to recognise as the achievement of liveability from a "whole of building" paradigm will not result unless assumptions associated with "good" design outcomes from the different professional perspective are clearly articulated and acknowledged. While apartment size and its inherent relationship with liveability was contested, functionality of space and liveability was not. Functionality of space was recognised as a viable alternative to prescribing minimum square metre requirements for apartments which can be easily manipulated through the inclusion of corridors and useless spaces.

Participant's responses identified that when considering liveability the relationship between the building and its neighbouring properties or public spaces is important. However, the focus was primarily at the boundary of the site and how that impacted the surrounding public realm. Little credence was given to the perceived value added in amenity uplift by large precinct developments. This is problematic given that increased densification is driven by the need to curb environmental impacts such as carbon emission in the face of growing populations through increased energy-efficient buildings that are supported by appropriate amenity. This notion of a building as a "precinct" is a new concept within the greater Melbourne CBD area. While the literature provides examples of successful precinct developments in Asia and the UK, evidenced in the work of [8], there is limited understanding around how they may translate in a metropolitan Melbourne context.

Buys et al. [8] and Buxton et al. [7] identified planning policies and regulations as being a critical influence in the resultant standards of liveability in urban areas. Victorian strategic planning policies articulate high-density living and increased population in urban areas. However, there is evidence to suggest the current form and functionality of the buildings recently developed are somewhat lacking in terms of their liveability [7, 9]. Despite the recently introduced apartments standards, there is limited policy guiding the provision of internal building amenity important in the context of increased building and city block density. Further, the quality (physical construction and environmental performance) of construction as informed by the Building Code of Australia was believed to only result in adequate outcomes. Given that noise and ventilation were recognised by respondents and in the conceptual framework as directly informing liveability, building standards are an important consideration. Additionally, it was identified that the novation of design, present in "design and construct" construction contracts, clearly presents issues around good design outcomes and the loss of liveability in the name of cost savings. What was interesting is that participants articulated that it was the financial institutions that pushed for such novation of design to ensure that one party could be held accountable. While the literature reviewed did not examine the impacts that banks and lending institutions have in the paradigm of liveability, from the results it can be seen that their influence may be significant and need to be considered in greater detail in the context of liveability.

The interviews recognised that there exists a level of disagreement across key industry stakeholder in terms of regulations and standards as they inform design outcomes and liveability. However, all stakeholder groups advocated the need to adopt a more collaborative approach. There is evidence in the findings of participants articulating their desire to work with other industry groups and do away with the stereotypically adversarial roles. A collaborative approach has the potential to benefit all parties, including the end users. Unfortunately, at present no such mechanism exists within the current legislation.

8 Conclusion

The performance of the urban form is not limited to simply the environmental performance of a building, but how it informs the social context of those that reside within it. Sustainable development is about the integration across environmental, social and economic paradigms. The increased densification of the urban form as a driver of improved environmental performance must consider the social and human capital that results from its manifestation. While there is a global trend to increase the number of people living at height, there is limited understanding of the impacts that this has on individuals and how decisions are made in relation to the product that is brought to market. The liveability of the urban form as it becomes denser will need to continue to be explored as it has increasingly become the way for life for the majority, especially in countries, like Australia, where it is not the norm.

Central to the evolution of regulations and processes that result in better "lived" experience for residents in high-rise apartment buildings is an understanding of what defines liveability.

While definitions of liveability are subjective, a number of key characteristics have been identified as contributing positively to a high level of liveability. These characteristics include: good apartment design that recognises the need for functional utilisation of space and facilitate a high quality indoor environment; building amenity that includes shared spaces within the building to allow for a diversity of uses that residents cannot undertake within their apartments, a sense of security and community; and consideration of external amenity, surrounding amenity that is required for the development of high levels of social and human capital. Such outcomes require an integrated approach with relevant professions working together to achieve an integrated approach. This is important given the research identified that the components of liveable are valued differently. This integrated approach reflects that planning a precinct and addressing the amenity are required for the proposed dwellings, not simply the dwellings in isolation. Current regulation, both planning and building, fails to overtly identify high-rise apartment buildings as precincts in themselves. In addition, it was recognised that there are indirect drivers of design outcomes that compromise liveable outcomes, specifically the role of financial intuitions and construction contracts. While the built form and its liveability has been questioned and debated, it is important to recognise that all stakeholder groups recognised were willing to participate in a more collaborative approach. However, how this is to be achieved; proactive, reactive regulation or market mechanisms such as a change in funding arrangements or financial incentives are topics of much debate among different stakeholders. These findings are relevant to both local and international housing market when considering how to improve the overall performance of the urban built form. Further research is required with a more extensive sample, to explore these key themes and determine the impact they have on the liveability of high-rise apartment developments.

References

1. Australian Bureau of Statistics (ABS) (2012) Victoria in future 2012 ABS. Canberra, Australia
2. Albrechts L (2010) More of the same is not enough! How could strategic spatial planning be instrumental in dealing with the challenges ahead? Environ Plann B: Plann and Des 37:1115–1127
3. Andrews CJ (2001) Analysing quality of place. Environ Plan 28:201–217
4. Baeten G (2002) The spaces of utopia and dystopia: introduction. Geogr Ann Ser B, Hum Geogr 84(3–4):141–142
5. Burke T, Stone M, Ralston L (2011) The slow burn of declining housing affordability and spatial polarisation. In: ENHR Conference, 5–8th July 2011, Toulouse, France
6. Buxton M (2013) This is not a plan. It is a hoax driven by money. The Age 10 Oct 2013, Viewed 15th Mar 2015, http://www.smh.com.au/comment/this-is-not-a-plan-it-is-a-hoax-driven-by-money-20131009-2v8hi.html#ixzz3463rcyQP
7. Buxton M, Goodman S, Moloney S (2016) Planning Melbourne: lessons for a sustainable city. CSIRO Publishing, Victoria, Australia
8. Buys L, Vine D, Miller E (2013) What makes inner city high density living attractive? Insight from residents in Brisbane, Australia. Environ Manage Sustain Dev 2(1):14–33
9. City of Melbourne (CoM) (2013) Future living: a discussion paper identifying issues and options for housing our community. City of Melbourne Research Division, Melbourne
10. Cotgrave A, Alkhaddar R (2006) Greening the curricula within construction programmes. J Educ Built Environ 1(1):3–29
11. Department of Environment, Land, Water and Planning (DWELP) (2017) Plan Melbourne 2017–2050 2017–2050 summary. Victorian Government, Victoria, Australia
12. Department of Planning, Transport, Local Government and Infrastructure (DPTLI) (2014) Plan Melbourne 2017–2050: a new metropolitan planning strategy. Victorian Government, Australia
13. Edwards B (1999) Sustainable architecture: European directives and building design, 2nd edn. Architectural Press, Oxford, pp xiv–xvi, 229
14. Forrest R, Kearns A (2000) Social capital, social cohesion and the neighbourhood. Urban Stud 38(12):2125–2143
15. Forsyth A (2003) Measuring density: working definitions for residential density and building intensity. Design centre for American urban landscape, viewed 29th Mar 2016. http://www.corridordevelopment.org/pdfs/from_MDC_Website/db9.pdf
16. Gifford R (1997) Environmental psychology. Allyn & Bacon, Boston

17. Hodyl L (2014) To investigate planning policies that deliver social outcomes in hyper-dense, high-rise residental environments. The Winston Churchill Memorial Trust of Australia, Australia
18. Howley P, Scott M, Redmond D (2009) Sustainability versus liveability: an investigation of neighbourhood satisfaction. J Environ Plann Manage 52(6):847–864
19. Hutchinson S, Skodol-Wilson H (1992) Validity threats in scheduled semi structured research interviews. Nurs Res 41(2):117–119
20. Landcom (2011) Residential density guide: for Landcom project teams. Sydney, Australia
21. Liu AMM (1999) Residential satisfaction in housing estates: a Hong Kong perspective. Autom Constr 8(4):511–524. http://dx.doi.org/10.1016%2FS0926-5805%2898%2900098-3
22. Martel A, Whitzman C, Fincher R, Lawther P, Woodcock I, Tucker D (2013) Getting to yes: overcoming barriers to affordable family friendly housing in inner Melbourne. State of Australian Cities Research Network, Sydney, Australia, 26–29 Nov 2013, viewed 10 Mar 2016. http://apo.org.au/resource/getting-yes-overcoming-barriers-affordable-family-friendly-housing-inner-melbourne
23. Major Cities Unit (2010) State of Australian Cities 2010. Infrastructure Australia, Canberra
24. McFarland P (2015) A tale of two cities: Sydney and Melbourne's growth strategies and the flawed city-centric approach, viewed 26th Mar 2016. http://www.soacconference.com.au/wp-content/uploads/2013/12/Mcfarland-Environment.pdf
25. Myers D (1988) Building knowledge about quality of life for urban planning. J Am Plann Assoc 54(3):347–358. https://doi.org/10.1080/01944368808976495
26. PIA (Planning Institute of Australia) (2003) Planning healthy communities: a PIA discussion paper on public health and wellbeing: a basis for developing a draft policy statement on planning for health. PIA, Victoria, Australia
27. Townsend M (2003) Prescribing parks for positive health outcomes. Plann News 29(3):5–6
28. Whyte WH (2001) The social life of small urban spaces. Project for Public Spaces, New York
29. Williamson R (1981) Adjustment to the highrise: variables in a German sample. Environ Behav 13(3):289–310
30. Wong G (2015) The only way to build is upwards: the tallest public housing development project in Singapore, viewed 6th Apr 2015. http://www.prres.net/Papers/Wong_The_Only_Way_to_Build_is_Upwards.pdf
31. Yeh AGO (2000) The planning and management of a better high density environment. In: Ng MK, Yeh AGO (eds) Planning for a better urban living environment in Asia. Ashgate, England, pp 116–143
32. Yeh A, Yuen B (2011) Tall building living in high-density cities: a comparison of Hong Kong and Singapore. In: Yeh A, Yuen B (eds) High-rise living in Asian cities. Springer, Amsterdam, pp 9–23

Dr. Sarah Holdsworth B.Ed. (Env. Sci.), M.Env.Sci., Ph.D. is Senior Lecturer in sustainability with expertise in sustainable education for the building industry, specifically notions of sustainability in relation to curriculum development and professional education.

David Kenny is a Graduate of the Bachelor of Applied Science (Construction Management) (Honours), RMIT University, Australia.

Jeremy Cooke is a Graduate of the Bachelor of Applied Science (Construction Management) (Honours), RMIT University, Australia.

Shaun Matfin is a Graduate of the Bachelor of Applied Science (Construction Management) (Honours), RMIT University, Australia.

The Way Forward—Moving Toward Net Zero Energy Standards

Adisa Adejare Alawode and Priyadarsini Rajagopalan

Abstract Net zero energy building standards have been gaining prominence lately as the next performance target for buildings. However, despite the demonstrated benefits of such building performance across triple bottom-line concepts, Australia is yet to formulate a policy toward adopting a net zero energy building standard. Evidence from various scholars suggests that Australia cannot delay the implementation of deep improvements in energy efficiency in the built environment any longer, as issues of energy security, affordability and increasing greenhouse gas emissions have become critical. This chapter reviews recent advances in the high-performance building standards with emphasis on global developments of net zero energy standards and discusses how Australia is positioned in relation to this standard and the ways Australia might move forward to this standard.

1 Introduction

The building sector has the potential to reduce its share of greenhouse gas (GHG) emissions by at least 30–35% while still accommodating growth in the overall number of buildings and population by 2050 [7]; see also Chapter "The Built Environment in Australia". Even though the issue of energy efficiency and reduction of GHG emissions across the built environment have dominated policy debates in Australia for some time, most of these efforts focused on limited performance targets and setting incremental steps toward better energy efficiency [31]. Current energy efficiency measures in the built environment have been unable to deliver the level of energy efficiency that Australia requires for a transition to a low-carbon future and

A. A. Alawode
School of Property Construction and Project Management, RMIT University, Melbourne, VIC, Australia

P. Rajagopalan (✉)
Sustainable Building Innovation Laboratory, School of Property Construction and Project Management, RMIT University, Melbourne, VIC, Australia
e-mail: priyadarsini.rajagopalan@rmit.edu.au

P. Rajagopalan et al. (eds.), *Energy Performance in the Australian Built Environment*, Green Energy and Technology, https://doi.org/10.1007/978-981-10-7880-4_13

are generally subpar when compared with other developed nations [20, 31, 40]. To combat this lack of progress, there is the need for a paradigm shift from what is deemed politically achievable to what the climate science demands to ensure the built environment achieves sustainability outcomes [31].

As discussed in Chapter "The Built Environment and Energy Efficiency in Australia: Current State of Play and Where to Next", the potential of the Australian built environment to achieve a low-energy and low-carbon future depends largely on raising the current minimum standards of building performance to global best practice levels. Although many of the current building rating systems promote energy reduction targets that minimize the environmental impacts of buildings, these targets are not sufficient to have a meaningful impact on climate change [15]. The need for higher standards has led to the interest in other standards that may bring deeper cuts in buildings' energy use. Some of the prominent building standards that aim to achieve deeper cuts in building energy use are: Passive House (Passivhaus) standards, nearly zero energy building standards (nZEB), and net zero energy building standards (NZEB). The chapter discusses the concept of highly energy-efficient buildings, the recent advances in those concepts, how Australia is positioned among other countries and the way forward to achieve higher standards.

2 Passivhaus (Passive House) Standards

Passivhaus (Passive House) is a globally recognized standard developed in Germany for the design of very low-energy buildings. This rigorous standard was developed in the early 1990s and is based on the idea of using simple, direct, and primarily architectural solutions to create ultra-low energy buildings. The standard started for housing, but over the years has expanded to commercial, industrial, and public buildings. Passive House standard employs strategies such as orientation, shading, passive solar gain, high envelope thermal performance, minimized thermal bridging, high levels of air tightness and mechanical ventilation with energy recovery to achieve its objective. The performance targets of the Passive House standard require the following criteria be met:

- Specific heating demand less or equal to 15 kWh/m^2.year (or)
- Specific heating load less or equal to 10 W/m^2
- Specific cooling demand less or equal to 15 kWh/m^2.year
- Specific primary energy demand less or equal to 120 kWh/m^2.year
- Air tightness less or equal to 0.6 ach at 50 Pa (n50).

Although the primary energy usage is very low in a Passive House, energy consumption is not equal to zero. Figure 1 shows the five principles of the standard which is applicable to residential, commercial, institutional, and industrial buildings both for new build and retrofitting of existing buildings. The Passive House Institute (PHI) is an independent research institute that has played a crucial role in the development of the Passive House concept and has assumed a leading position with

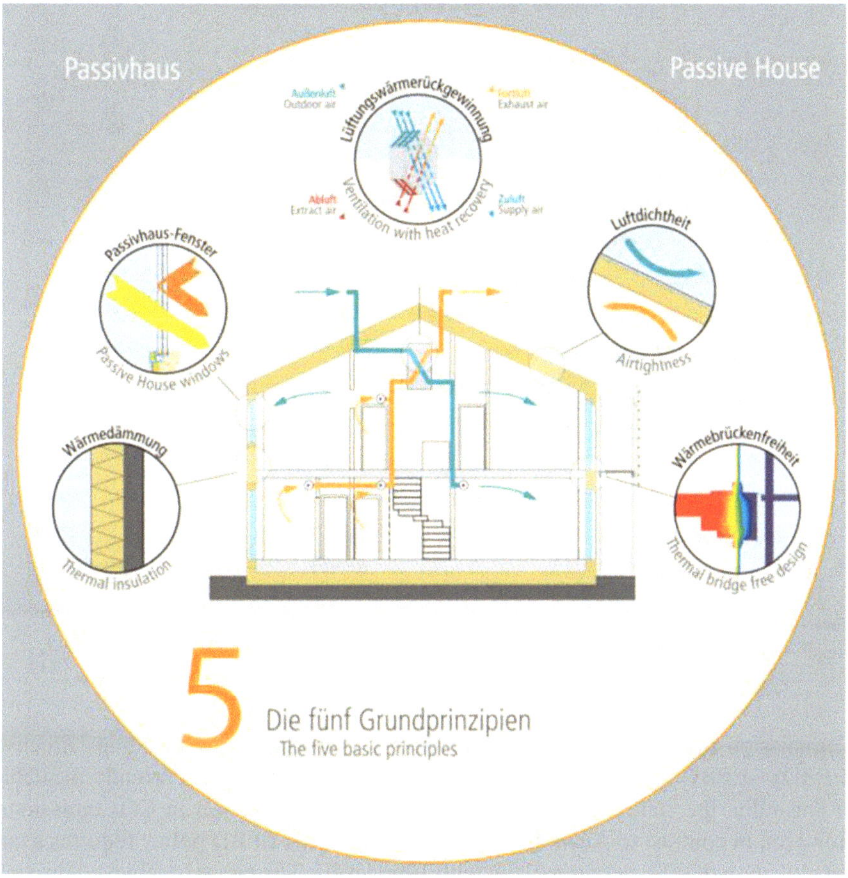

Fig. 1 Five principles of Passivhaus standards (*Source* [35])

regard to research on and development of construction concepts, building components, planning tools, and quality assurance in this area. As of the time of preparing this manuscript, there are 1195 certified Passive House buildings globally of which 6 are in Australia [35].

3 Nearly Zero Energy Building (NZEB) Standards

A nearly zero energy building (nZEB) refers to building with very low energy requirement; a significant portion of the energy required to operate the building is drawn from renewable sources produced on site or nearby [14, 25]. Figure 2 is a graphical representation of the definition of an nZEB which includes renewable energy source outside the building site. nZEB as a standard was mandated by the European

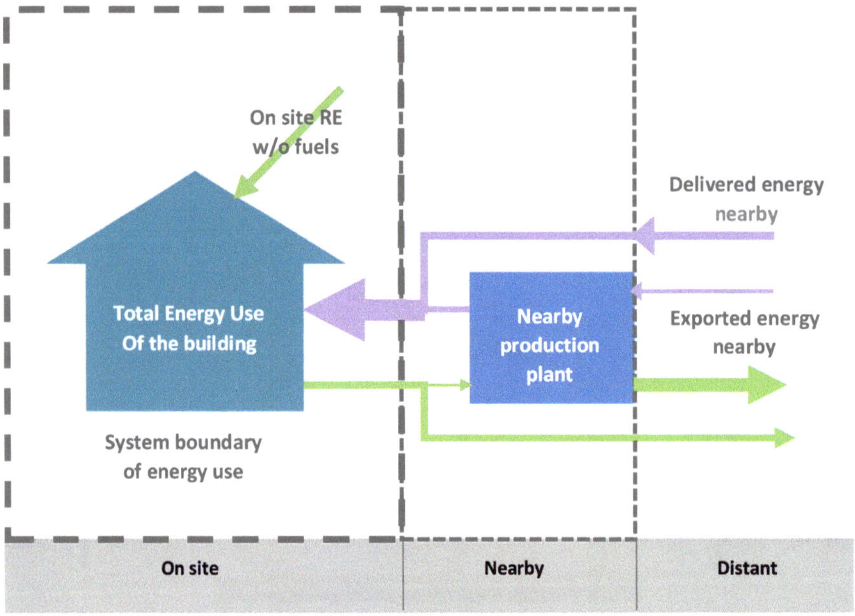

Fig. 2 Graphical definition of nZEB standard (*Source* [25])

Commission through the recast of the Energy Performance of Building Directive (EPBD)—EPBD/2010/31/EU. Similar to environmental targets currently available in Australia, the standard was initiated to achieve a reduction in CO_2 emissions. However, in contrast to Australian emission targets, the EPBD policy requires a target reduction of 90% in comparison with 1990 levels. The target is achieved through improving energy efficiency, reducing energy consumption, and increasing the share of renewable energy in the building sectors of Member States [5]. The policy also stipulates that all new buildings across the EU achieve the nearly Zero Energy (nZE) goal by the year 2020. Similarly, all existing buildings are expected to achieve the target by the year 2050 [14]. The directive allows EU Member States to work out their individual plan of achieving the goal such as determining definitions and specific building requirements [14]. More than a half of the Member States already implemented a definition considering the share of renewables in a quantitative or qualitative way, and some are under approval. Several Member States provided a definition. Other Member States have a definition under development [14].

The design of nZEB building focuses on the use of passive design principles such as envelope insulation, daylighting, natural ventilation and evaporative cooling, and also the use of high-efficiency HVAC systems [5].

4 Net Zero Energy Building (NZEB) Standards

Net zero energy building (NZEB) standards have gained popularity in recent years as the next target for building's energy performance globally. Various literature have associated this popularity to the contributions of NZEB standards to addressing key national and global issues such as energy security, GHG emissions, and better quality indoor environments [2, 6, 8, 27–29, 33]. As such, there is a general agreement among building scholars that NZEB standards for buildings hold significant potential in helping the construction sector achieve its share of environmental targets, including better-performing buildings. In concept, a NZEB is a very low-energy building that balances its low annual energy consumption by the use of renewable energy on site. Depending on the variables being considered, for example, energy metric, building boundary, context, or calculation methodologies, NZEB have been defined in several ways [42, 39]. Most definitions agree that a NZEB will be highly efficient and make use of renewable energy extensively. Of all the definition variants available in the literature, the definition proposed by Torcellini et al. [42] of the National Renewable Energy Laboratory based on the supply-side options available on site is more commonly referenced in the literature. The four definition options as proposed by Torcellini et al [42, p. 5] are:

- Net Zero Site Energy—"a ZEB that produces as much energy as it uses in a year when accounted for at the site".
- Net Zero Source Energy—"a ZEB that produces at least as much energy as it uses in a year, when accounted for at the source".
- Net Zero Energy costs—"a ZEB where the amount of money the utility pays the building owner for the energy the building exports to the grid is equal to the amount the owner pays the utility for the energy services and energy used over the year".
- Net Zero Energy Emissions—"a ZEB which produces at least as much emissions-free renewable energy as it uses from emissions-producing energy sources".

While each definition serves NZEB based on the different criteria, a publication by Keeler [23] suggested there are wide differences and that each of the definition classes achieves the NZE goal in different ways. Table 1 is a summary of the pros and cons of each of the definition classes as suggested by Keeler [23].

Scholars have suggested Australia's immediate adoption of NZEB standards as the next performance target for buildings [9, 17, 24]. Some of the OECD countries including the EU, Japan, South Korea, UK, USA, and Canada already have in place-specific policy that addresses the adoption of net zero energy/carbon standards in their building sector. For example, the Korean Government initiated a policy in 2009—"Measures to Develop Green Cities and Buildings" with specific energy efficiency targets including that all residential buildings should achieve zero energy/emissions by 2025 [18]. Japan, via the "Basic Energy Plan" policy, targets to have all newly constructed public buildings be net zero energy by 2020 and zero energy by 2030 [32]. Similarly, The USA, via the "Energy Independence and Security Act of 2007, Title IV, Subtitle B, Section 422" established a zero energy policy that has as its goal, the achieve-

Table 1 NZEB definition classes, pros and cons (*Source* [23])

			Renewable Energy location option			
			NZEB A	NZEB B	NZEB C	NZEB D
NZE type	Pros	Cons	Building Footprint	Site Boundary	Off-site	RECs
Site Energy	Energy measured on site with meters and utility bills Easy to understand and communicate Encourages energy-efficient building design	Requires more on-site electricity to offset natural gas use Does not differentiate between fuel types for their emissions or other environmental impacts Values energy import and exports equally, not accounting for additional export costs	Difficult Limited areas for energy generation, harvesting, biomass is unlikely	Possible More area for generation and biomass	Easy More area for generation and biomass	Easy No geographic restriction on acquiring renewable energy credits (RECs)
Source Energy	Easier NZE goal to reach due to on-site energy generation being accorded a higher value Includes fuel distribution and generation impacts Values each fuel used at site differently	Easy to implement Does not differentiate between fuel types for their emissions or other environmental impact Site-to-source conversions needed Not easy to understand and communicate	Possible When on-site renewable energy generation is values higher	Possible When on-site renewable energy generation is values higher	Difficult When biomass is used in large quantities	Easy No geographic restriction on acquiring RECs
Energy Cost	Easy to implement and measure Values and allows demand response control Verifiable from utility bills	Volatile energy rates make prediction of performance and comparison from year to year difficult	Difficult Depends on utility purchase rates	Difficult When biomass is used in large quantities	Difficult When biomass is used in large quantities	Possible If RECs are cheap, or bought in large quantities
Emissions	Easier NZE goal to reach Accounts for non-energy differences between fuel types such as pollution and greenhouse gases	Accounting is dependent on emissions information from utilities, which is likely to be historical data	Possible When on site generation has favorable emission factors	Difficult When biomass is used in large quantities	Difficult When biomass is used in large quantities	Possible When the RECs have tolerable emission factors

ment of zero-net energy for all new commercial buildings built after 2025 and for all pre 2025 buildings to be net zero energy by 2050 [41]. However, a deliberate government-led policy program for Australia to transition to zero energy standards is currently missing [17, 44]. While different factors (market and others) could be attributed to this lack of or slow adoption, a failure to act cannot be excused [24], as the addition of new buildings performing below NZE levels will likely pose severe strain on the existing energy supply infrastructure and will also impact Australia's ability to meet its commitment to global efforts at curbing climate change [9].

5 Defining Net Zero

Since Australia is yet to develop a firm policy on NZEB, the question of how to define this outcome or which of the existing definitions to adopt becomes very crucial and of immediate importance. This is because a lack of clear definition will not only hamper policy but will also affect how strategies by which the NZE standard can be achieved [34]. Care should be taken to ensure the choice or choices of definitions address the strategic interests of all levels of government and also consider other market players. These interests should be properly aligned so that the ensuing definition can be clear, consistent, and deployable; this is a necessary first step for Australia to transition to NZE performance standards. Australia needs to develop a policy framework that supports higher energy efficiency and deeper cuts in energy use in the building industry, as it is only when such policy environment exists—as different from the current policy that seems to only encourage incremental performance standards—will Australia be able to transition NZE performance standards. Such a framework will need to address among other issues, how the strategic interests of the different state actors and the entire supply chain in the industry will be aligned. Australia can learn from the successes and shortcomings of the NZE policy of other OECD countries which are discussed in the next section.

5.1 Net Zero Energy Building Policies: Lessons from Other Countries

Some OECD countries have led the way in addressing the NZEB challenges, and Australia can learn lessons from their implementation of such policies to support the growth of the NZEB standard locally. This section presents a brief overview of NZEB policies in some of the OECD countries.

A number of policy programmes such as: The 2030 Challenge and The California Big Bold Energy Efficiency Strategy (BBEES) are currently in place in the USA to address NZEB and its widespread adoption in the market. Many of the initiatives currently in place came from not just the government but also the private sector.

5.1.1 The 2030 Challenge

The 2030 Challenge was issued in 2005 by Architecture 2030—a non-governmental organization established in the USA by Edward Mazria. The 2030 challenge aims to challenge design teams globally to immediately implement a 60% reduction in fossil fuel-based energy consumption for new and renovated buildings and infrastructure. The challenge calls for the building industry to cut CO_2 emissions by 70% by 2015, 80% in 2020, and 90% in 2025 and to be carbon neutral by 2030. The targets allow for the use of innovative sustainable design strategies, generating on site renewable energy and/or purchasing up to 20% renewable energy from outside organizations. Part of the strategies adopted by the organization in encouraging participation are outreach programs to governments at all levels, professional development programs, and provision of free design tools targeted at industry practitioners. The Challenge has been adopted and being implemented by a large number of architecture/engineering and planning firms in the USA including organizations like American Institute of Architecture (AIA), American Society for Heating Refrigeration and Air conditioning (ASHRAE), the US conference of Mayors, federal, state, and local government agencies. The 2030 Challenge formed a significant part of the mandate issued for federal buildings in the "Energy Independence and Security Act," and by the end of 2013, there were established 2030 districts in Denver, Seattle, Cleveland, Pittsburgh, and Los Angeles representing 107 property owners, 111 professional and community stakeholders and over 9 million square meters of committed real estate. Reports from the US "2016 Annual energy Outlook" in January 2007 confirms a reduction in the energy consumption projections for the building sector (building operations) to the tune of 18.5 Quadrillion BTUs since 2005 [12].

5.1.2 The California BBEES

Following the California assembly bill 32 (AB 32), the California Global Warming Solutions Act 2006 which requires that greenhouse gas emissions in the state be reduced to 1990 levels by 2020, the California Public Utilities Commission adopted the California Long Term Energy Efficiency Strategic Plan (CEESP) which created a roadmap for scaling up state-wide energy efficiency measures to sustain market transformation. The strategic plan has four programmatic components called Big, Bold, Energy Efficiency Strategies (BBEES) which aimed to not only improve energy efficiency, but also to galvanize market players. The 2013 Integrated Energy Policy Report [10] provided further insights including definition for NZEB and also allowed for "development entitlements" for off-site renewable energy sources as a viable option for builders and developers. The report also discussed the integration of the NZEB performance requirements into California building standards. Table 2 shows the components of California strategic plan. At the federal level, legislations such as the Energy Policy Act of 2005 and Energy Independence and Security Act (EISA) of 2007 exist and were designed to move the market toward NZEB. For example, Section 423 of the EISA mandated the US department of Energy (DOE) to establish

Table 2 California BBEES (adapted from: California Long Term Energy Efficiency Strategic Plan Report, 2008)

Market Sector	Program goal
New Residential	All new residential construction will be net zero energy by 2030
New Commercial	All new commercial construction will be net zero energy by 2030
HVAC industry	Industry will be transformed to ensure energy performance is optimal for California climate
Low-income Customers	Eligible low-income customers will be given the opportunity to participate in low-income energy efficiency program by 2020

a national clearing house to provide information and public outreach about high performing buildings. In addition, there are also a number of non-R&D financial incentives in the form of loans, grants, corporate deductions, corporate exemptions, and personal tax exemption programs aimed at encouraging uptake of NZEBs.

As discussed earlier, the EPBD recast of 2010 mandated all Member States within the European Union to adopt the nearly zero energy building standards in their individual built environments. Apart from setting broad targets and dates for attaining the standards for buildings, the EPBD also recognized the role played by cost, improving the potential for a successful deployment. Part of the expectations from Member States therefore includes cost-optimal solutions for achieving the standard. The Power House report [37] reported on strategies that have been adopted by some of the Member States toward encouraging nZEB. Some of these approaches include research and development programs, voluntary standards, promotion schemes and financial incentives and intense exchange of experience as well as training and networking.

5.2 *Achieving Net Zero Energy Goals*

There are many strategies or pathways to achieve the NZEB goal. Aelenei et al. [4] identified three approaches that dominate most of the pathways as: Passive Design Strategies (PDS), the goal of which is to reduce energy demand; use of high-efficiency energy systems; and the use of renewable energy to offset the energy demand and reach the NZE goal [4]. Further elaboration of these approaches was documented in works such as Garde et al. [16], Cellura et al. [11], Kwan and Guan [26], and Wang et al. [43]. Figure 3 shows a graphical representation of the components of the NZEB design.

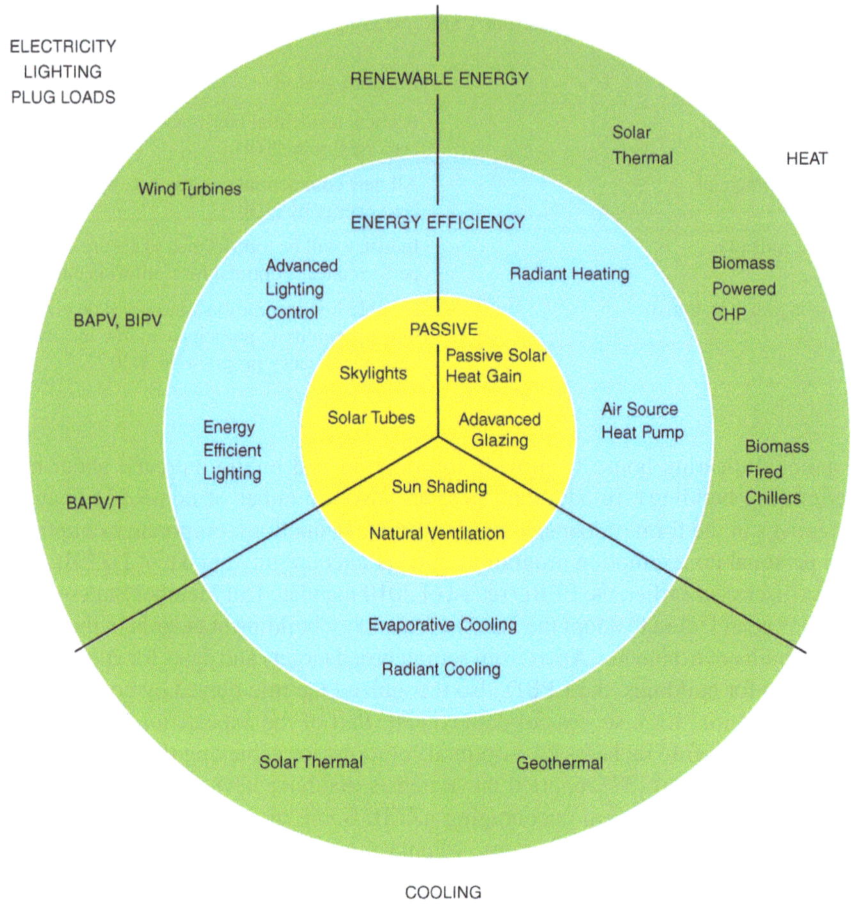

Fig. 3 Design of NZEB (*Source* [3])

5.3 *Passive Design Strategies*

For most buildings, a large part of energy consumption is related to the use of active systems to maintain a comfortable indoor environment. As such, Passive Design Strategies (PDS) are considered an important approach to improving the energy performance of buildings and to achieve indoor comfort without relying heavily on active HVAC systems [38]. Passive design is defined as "the use of architecture and climate to provide heating cooling, ventilation and lighting." It is also regarded as the use of architecture to harvest free energy from the environment' [19, p. 185].

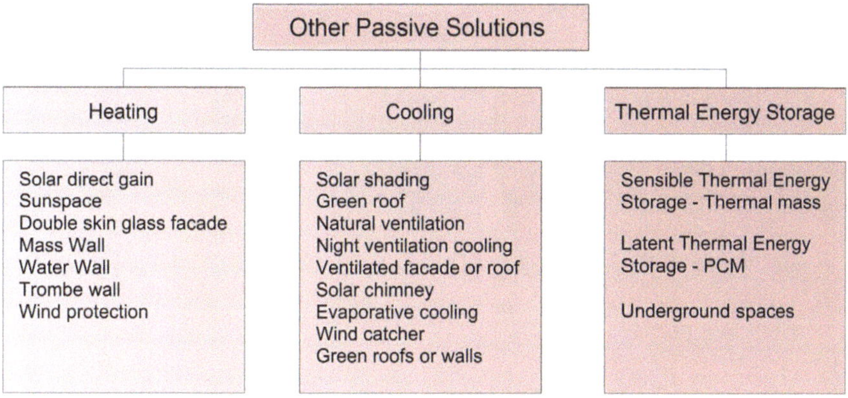

Fig. 4 Passive design solutions for net zero energy buildings (*Source* [38])

As Passive Design Strategies are specific to climates, building orientation to achieve solar heat gain or prevent heat gain in the building fabric, ventilation, and envelope solutions play major roles. Rodriguez-Ubinas [38] documented some of the Passive Design Strategies that may be employed for the design of NZEB as shown in Fig. 4.

5.4 *Active Systems*

A large proportion of building's energy use goes into running active systems. Energy efficiency therefore plays an important role in ensuring that these energy end-use sources operate at their optimum as this is crucial to the achievement of the NZEB goal [4]. Abdellah et al. [1] suggested that energy efficiency can be improved by employing different strategies including mechanical ventilation with heat recovery systems, and the use of high-efficiency technologies such as low power lighting and energy-efficient (high-star-rated) appliances. Figure 5 shows some of the active systems as suggested by Aelenei et al. [3].

5.5 *Renewable Energy Sources*

A net zero energy building requires intensive use of renewable energy resources to achieve "zero" use of fossil fuels [13, pp. 634–635, 21]. Renewable energy supply is therefore considered the third major component of any NZEB. McCrea [30, p. 9] defines renewable energy as those "obtained from sources that are inexhaustible, emit no greenhouse gases or are emission neutral over their lifetimes. They are

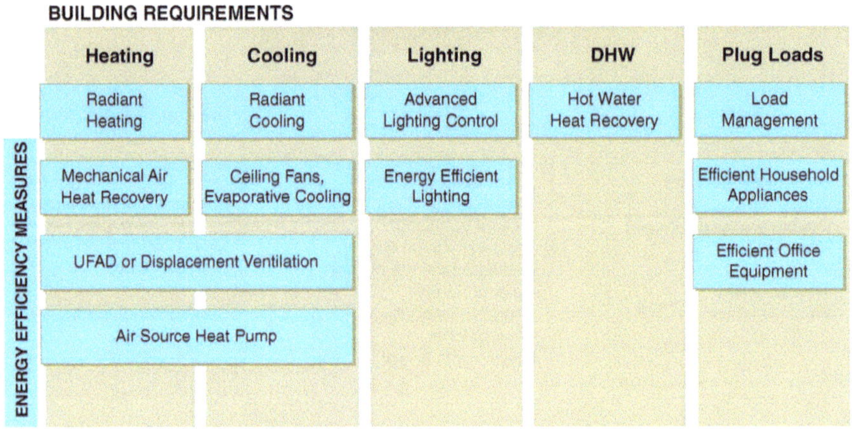

Fig. 5 Energy efficiency measures for net zero energy buildings (*Source* [3])

also energy sources replenished by natural phenomena." Figure 5 shows some of the systems/technologies available in the Australian market including solar thermal, solar photovoltaic, and wind-based renewable energy systems.

6 Discussion

NZEB standards have the potential to address key national and global issues. With advances in construction technologies and renewable energy systems, creating NZEB is becoming more feasible. Though the Australian Government has made efforts to incrementally improve the energy performance of buildings, the current energy performance standards have not reached the level of comparable developed economies. The absence of appropriate policy initiatives, targets, and standards relating to developing and encouraging NZEB may severely affect Australia's ability to meet its commitment to global efforts at curbing climate change. Owing to the different levels of governing in Australia, challenges may exist both with the effectiveness of policy making and implementation toward NZEB. However, the challenges pale in significance when compared to the benefits that can be reaped when Australia transitions to a standard that encourages deeper energy cuts like NZEB. Moreover, Australia can learn policy and implementation lessons from other OECD countries some of whom already have established NZEB policies that are already entrenched in their building codes and regulations and can therefore overcome many of these challenges.

Moving forward, a successful transition to NZEB standard will involve various considerations, as suggested by the International Energy Agency's Solar Heating and Cooling (IEA SHC) Task force 40 [21], part of the considerations to achieve widespread NZEB will be the need for market-oriented initiatives that include incentives to encourage adoption of NZEB standards and technology and support NZEB marketing activities. Mandatory disclosure of energy performance has helped the commercial buildings to move in the right direction. In the case of residential buildings, the implementation is straightforward for detached houses, but more challenging for multi-residential buildings as they may not appear cost-effective for building developers and owners. Although a recent report by Pitt & Sherry [36] suggests that it is now both technically and commercially feasible for high-rise residential buildings to reach the NZEB target in the Australian market, there is the need to investigate further to verify the feasibility and document successful strategies for wider applications in the building industry. Most of the publications on achieving NZEBs have focused mostly on technical issues. It is important to "monetize the social benefits" of net zero, otherwise the transition to net zero buildings will be glacial [22]. However, the wider societal benefits of net zero are not being appropriately valued due to the policy gaps and barriers. Cost is a key variable that drives most real estate development projects; it is therefore important to consider how NZEB projects can be cost competitive if such projects are to be adopted by the market. A lack of market transformation strategies for high-performance building designs and materials keeps key elements like high-performance glazing in high-cost, niche markets [22]. In addition, electricity network pricing arrangements do not fully reward the developer for avoiding the need to enhance network capacity, or for reducing peak demand.

7 Conclusion

The energy efficiency efforts currently pursued in Australia need to be scaled up to match global best practice standards. One of such standards is the NZEB which has been demonstrated to have both economic and environmental benefits if adopted. Given the lack of a holistic definition, determining best practices becomes difficult. Some OECD countries have led the way and already have a policy in place to support the adoption of the standard, and Australia have a large pool of resources to tap into by learning lessons from some of the policy initiatives of those other OECD nations. Higher level of code compliance in addition to cost reduction of high-end products is essential to move toward NZEB standards.

References

1. Abdellah RH, Nasid MA, Chen GK, Mohamed S, Omar R (2017) The potential of net zero energy buildings (nzebs) concept at the design stage of healthcare buildings towards sustainable development. IOP Conference series: Materials Science and Engineering, p 271
2. Adhikari RS, Aste N, Pero CD, Manfren M (2012) Net zero energy buildings: expense or investment? Energy Procedia 14:1331–1336
3. Aelenei L, Aelenei D, Cubi E, Musall E, Tae Kim J, Shaan C, Donn M (2017) Net zero energy building design fundamentals. In: Garde F, Ayoub J, Aelenei L, Aelenei D, Scognamiglio A

(eds) Solution sets for net zero energy buildings: feedback from 30 net zebs worldwide. Ernst & Sohn, Berlin
4. Aelenei L, Aelenei D, Gonçalves H, Lollini R, Musall E, Scognamiglio A, Cubi E, Noguchi M (2013) Design issues for net zero-energy buildings. Open House Int 38:7–14
5. Agostino D, Zangheri P, Castellazzi L (2017) Towards nearly zero energy buildings in europe: a focus on retrofit in non-residential buildings. Energies 10:117
6. Aksamija A (2016) Regenerative design and adaptive reuse of existing commercial buildings for net-zero energy use. Sustain Cities Soc 27:185–195
7. Asbec (2007) Capitalising on the building sector's potential to lessen the costs of a broad based ghg emissions cut. In: Sydney, C. F. I. E. C. (ed.). ASBEC, Canberra
8. Berry S, Davidson K, Saman W (2014) Defining zero carbon and zero energy homes from a performance-based regulatory perspective. Energ Effi 7:303–322
9. Berry S, Marker T (2015) Residential energy efficiency standards in australia: where to next? Energ Effi 8:963–974
10. California Energy Commission (2013) 2013 integrated energy policy report. California Energy Commission, California
11. Cellura M, Campanella L, Ciulla G, Guarino F, Brano VL, Cesarini DN, Orioli A (2011) A net zero energy building in Italy: design studies to reach the net zero energy target. In: Proceedings of building simulation, pp 649–655
12. Challenge (2017) Us building sector emissions down- the driving force: You! In: 2030, A. (ed)
13. Da Graça GC, Augusto A, Lerer MM (2012) Solar powered net zero energy houses for southern europe: feasibility study. Sol Energy 86:634–646
14. European Commission (2012) Towards nearly zero energy buildings: definitions of common principles under the epbd-final report. In: Ecofys (ed). European Commission, Germany
15. FXFOWLE Architects, Dagher Engineering, Passive House academy, Dharam Consulting, Simpson Gumpertz Heger (2017) Feasibility study to implement the passivhaus standard on tall residential buildings. New York State Research and Development Authority
16. Garde F, Lenoir A, Scognamiglio A, Aelenei D, Waldren D, Rostvik HN, Ayoub J, Aelenei L, Donn M, Tardif M, Cory S (2014) Design of net zero energy buildings: feedback from international projects. In: The 6th International conference on Applied Energy—ICAE2014. Energy Procedia
17. Gettler L (2014) Net zero energy buildings: why Australia must get a move on. The Fifth Estate [Online]. Available: http://www.thefifthestate.com.au/innovation/design/net-zero-buildings-why-australia-must-get-a-move-on/58346. Accessed 28 Feb 2017
18. Heechan K (2010) Korean green building code. SERI Q 3(105–111):8
19. Hootman T (2013) Integrated process. In: Net zero energy design: a guide for commercial architecture. Wiley, Hoboken
20. Horne R, Hayles C, Hes D, Jensen C, Opray L, Wakefield R, Wasiluk K (2005) International comparison of building energy performance standards. Report to Australian Greenhouse Office, Department of Environment and Heritage
21. Iea Shc Task 40 (2015) Towards net zero energy solar buildings-shc position paper. In: 52, E. A. (ed). International Energy Agency Solar Heating & Cooling Programme
22. Jewell C (2016) We can get net zero high-rise resi-with the right support. The Fifth Estate. https://www.thefifthestate.com.au/innovation/building-construction/we-can-get-net-zero-high-rise-resi-with-the-right-support. Last Accessed 18 Jan 2018
23. Keeler M (2016) Net zero energy and renewable energy systems. In: Vaidya P (ed) Fundamentals of integrated design for sustainable building, 2nd edn. Wiley, Hoboken
24. Kibert CJ, Fard MM (2012) Differentiating among low-energy, low-carbon and net-zero-energy building strategies for policy formulation. Build Res Inf 40:625–637
25. Kurnitski J (2013) Nearly zero-energy building's (nzeb) definitions and assessment boundaries. In: Kurnitski J (ed) Cost optimal and nearly zero-energy buildings (nzeb): definitions, calculation principles and case studies. Springer, London
26. Kwan Y, Guan L (2015) Design a zero energy house in Brisbane, Australia. In: 9th International Symposium on Heating, Ventilation and Air conditioning (ISHVAC) and the 3rd International

Conference on Building Energy and Environment (COBEE), 2015. Elsevier Science Direct, pp 604–611
27. Kylili A, Fokaides PA (2015) European smart cities: the role of zero energy buildings. Sustain Cities Soc 15:86–95
28. Li DH, Yang L, Lam JC (2013) Zero energy buildings and sustainable development implications—a review. Energy 54:1–10
29. Marszal AJ, Heiselberg P (2011) Life cycle cost analysis of a multi-storey residential net zero energy building in denmark. Energy 36:5600–5609
30. Mccrea A (2013) Renewable energy. Ramsbury, Marlborough, Wiltshire SN8 2HR, The Cromwood Press Ltd
31. Melbourne Energy Institute (2013) Zero carbon Australia buildings plan. Kindness House, Suite 10 Level 1, 288 Brunswich Street, Fitzroy, VIC 3065, Beyond Zero Emissions
32. Ministry of Economy Trade and Industry (2015) Meti compiles zeh roadmap. In: Ministry of Economy Trade and Industry (ed). Ministry of Economy Trade and Industry, Tokyo
33. Moore T (2014) Modelling the through-life costs and benefits of detached zero (net) energy housing in Melbourne, Australia. Energy Build 70:463–471
34. Nicholas RB, William MC, Roland RJ (2013) The prospect of zero net energy buildings in the United States. In: Sioshansi FP (ed) Energy efficiency: towards the end of demand growth. Academic Press, San Diego
35. Passive House Institute 2018
36. Pitt&Sherry (2016) Accelerating net-zero high-rise residential buildings in Australia-final report. City of Sydney
37. Powerhouse (2013) Nearly zero energy buildings overview: status, cost optimality and key conditions for nzeb. POWER HOUSE nearly Zero Energy Challenge Report
38. Rodriguez-Ubinas E, Montero C, Porteros M, Vega S, Navarro I, Castillo-Cagigal M, Matallanas E, Gutiérrez A (2014) Passive design strategies and performance of net energy plus houses. Energy Build 83:10–22
39. Sartori I, Napolitano A, Voss K (2011) Net zero energy buildings: a consistent definition framework. Energy & Buildings 48:220–232
40. Schimschar S, Blok K, Boermans T, Hermelink A (2011) Germany's path towards nearly zero-energy buildings—enabling the greenhouse gas mitigation potential in the building stock. Energy Policy 39:3346–4215
41. Sissine F (2007) Crs report for congress—energy independence and security act of 2007: a summary of major provisions. In: Service, C. R. (ed). CRS, Energy Policy Resources Science and Industry division
42. Torcellini P, Pless S, Deru M, Crawley D (2006) Zero energy buildings: a critical look at the definition. National Renewable Energy Laboratory and Department of Energy, US
43. Wang L, Gwilliam J, Jones P (2009) Case study of zero energy house design in UK. Energy Build 41:1215–1222
44. Wells L, Rismanchi B, Aye L (2017) A review of net zero energy buildings with reflections on the Australian context. Energy Build

Adisa Adejare Alawode is a Ph.D. student working in the area of net zero energy buildings.

Associate Professor Priyadarsini Rajagopalan, B.Tech. Arch. Eng., M.Bld.Sc., Ph.D. is an associate professor with expertise in building energy efficiency and energy benchmarking, indoor environmental performance, and urban microclimate.

Cohesion: Our Environment—Building Better and Smarter

Trivess Moore, Mary Myla Andamon and Priyadarsini Rajagopalan

The world is rapidly changing. Climate change is recognised as one of the greatest challenges facing the world today, and Australia is not an exception (see Chapter "Urban Climates in the Transformation of Australian Cities"). Environmental degradation is mainly due to anthropogenic greenhouse gas emissions. We are already seeing changes to weather patterns and more extreme and frequent weather events. This impacts on our built environment, our cities and our way of life. For example, hotter weather causes rising electricity demand due to increased requirements for cooling in buildings. A changing climate is also creating significant health and well-being challenges, especially during extreme weather events. The built environment is a significant contributor to rising greenhouse gas emissions. Increasing energy demand over recent decades, mostly from fossil fuel, needs to be addressed if the built environment is to transition to a low carbon and sustainable future.

The simple relationship between the design of buildings and climate can at times be set aside when creating pleasant internal environments [6]. Though just as important, the different variables that contribute to providing these internal environments have implications beyond the buildings themselves (see Chapters "The Built Environment and Energy Efficiency in Australia: Current State of Play and Where to Next", "Environmental Rating Systems for Non-residential Buildings—How Does Australia Compare with International Best Practice?" and the case studies presented in Part III). From the preceding chapter discussions, we appreciate now that this pervasive objective of having pleasant built environments affects the already changing climate.

T. Moore · M. M. Andamon · P. Rajagopalan (✉)
Sustainable Building Innovation Laboratory, School of Property Construction and Project Management, RMIT University, Melbourne, VIC, Australia
e-mail: priyadarsini.rajagopalan@rmit.edu.au

T. Moore
e-mail: trivess.moore@rmit.edu.au

M. M. Andamon
e-mail: mary.andamon@rmit.edu.au

P. Rajagopalan et al. (eds.), *Energy Performance in the Australian Built Environment*, Green Energy and Technology, https://doi.org/10.1007/978-981-10-7880-4_14

The built environment has been identified as having some of the best opportunities for reducing environmental impacts. This is through key strategies such as improving energy efficiency and the use of renewable energy. To achieve the type of low carbon outcomes required from the built environment, it will take a significant effort from all stakeholders. There is also a need for improved discussion and presentation of different attempts at improving energy performance so that we can all learn from what works, and what does not.

This edited book presents different elements of the energy performance in the built environment in Australia. In this book, there are many case studies of various building types covering residential, multi-residential, school and university buildings, as well as sports buildings. Together with the technical aspects of sustainable development, the integration across environmental, social and economic paradigms is highlighted in this book. There are also several chapters which discuss the regulatory approaches to addressing energy performance in buildings.

Similar to many other countries, various levels of government in Australia have attempted to address energy consumption in buildings through a variety of direct and indirect mechanisms and mandatory regulations and voluntary approaches. Energy programs designed to improve energy efficiency in buildings have had a significant impact on the Australian built environment. For example, since 1990, the average energy consumption per Australian household has remained relatively constant despite the increases in the average size of houses, the use of space conditioning equipment, and the diverse range of appliances now in every house [4]. The decline in energy consumption per household in terms of floor area is primarily being driven by improved efficiency of appliances and the building shell [4]. However, there is a need to understand occupants and changing housing needs to push towards a more sustainable built environment.

There are a number of opportunities for policy improvement which would help occupants as well as guide the building industry. Although local councils in Australia have made efforts to provide energy efficient strategies to reduce carbon emissions, there seem to be less benchmarks set for multi-residential buildings. A study by New South Wales Department of Infrastructure, Planning and Natural Resources (cited in [2]) showed that residents of high and medium density apartments consume 25% more energy than detached houses due to common areas such as foyers, corridors, pools, gyms and car parks. Further research on energy efficiency benchmarks to manage energy consumption in multi-residential buildings including common spaces is necessary. In addition, further analysis of energy consumption from a socio-economic perspective is required to fully understand the residents and to provide better design strategies in order to achieve building energy efficiency in apartments. Social housing presents additional challenges such as the ongoing maintenance costs for sustainability technologies such as solar photovoltaics and rainwater tanks for the providers. To make sustainable housing more affordable for social housing providers and for improving quality of life for tenants, more innovative ways to recoup some of the sustainability costs may be needed.

In the case of commercial buildings, the adoption rate of voluntary rating systems for new buildings such as Green Star has been limited to high-end office buildings in the CBD area of major cities. However, mandatory regulations like the Commercial Building Disclosure Programme seem to be helping to push the market. Verified base building NABERS energy ratings are capable of motivating all the supply side players and moving the market. The reluctance of tenants to occupy new space without knowing the occupancy rating has forced developers and investors to match measured building performance in line with design prediction [3]. International comparative studies note that Melbourne's best performing buildings are using three times less energy on a like-for-like basis compared to London's best performing new buildings [3]. Pushing the system harder makes innovation flourish, and this is no doubt beneficial to all stakeholders. Cutting edge new buildings at the 6 star Green Star performance level lead to the development of nearly zero energy buildings.

The positive outlook for distributed PV adoption as a result of unprecedented cost reductions experienced among photovoltaic and electricity conversion technologies is evident. Studies indicate that the deployment of solar PV is growing in commercial buildings, with a number of businesses across the country having now installed a solar power system, in addition to more than 1.7 million residential installations [1]. However, the absence of appropriate standards relating to Net Zero Energy Buildings will severely affect Australia's ability to meet its commitment to global efforts at curbing climate change. The wider societal benefits of net zero are not being appropriately valued due to the policy gaps and barriers. In addition, electricity network pricing arrangements do not fully reward developers in Australia for avoiding the need to enhance network capacity, or for reducing peak demand.

While the required indoor environmental performance for office workers is well understood, there is clear absence of documentation on the state of indoor environments in educational facilities particularly for schools and early childhood centres. Therefore, a study involving measurements and surveys of temperature, comfort conditions, and the relationship between these aspects of indoor environments and student performance should be undertaken urgently. Nonetheless, it is encouraging to see key universities engaging more seriously with improving sustainability outcomes demonstrating it to the hundreds of thousands of students who use these facilities, not only in Australia but globally.

We called this book *Energy Performance in the Australian Built Environment* with emphasis on the impact of built environments. However, the idea that environmental issues are not distinct from social ones also underpinned much of the discussions in the chapters. In the Paris Agreement, the year 2020 is expected to be the world's deadline for putting greenhouse gas emissions on a downward path if we are to have any chance of keeping climate change in check. As debates continue over 2 °C degrees versus 1.5 °C, climate adaptation versus mitigation and the necessity of building smarter built environments, we are reminded of Richard Rogers' take in the *City for a Small Planet* [5], "*policies aimed at improving the environment can also improve the social life of citizens*" (p. 32).

To conclude, it is important to recognise that all stakeholders should participate in a more collaborative approach to improve the energy performance of our built environment. The best methods of achieving this, such as proactive approaches, mandatory regulations or market mechanisms are topics for ongoing debate.

References

1. ARENA (2018) Renewable energy in residential & commercial buildings. Australian Government Australian Renewable Energy Agency (ARENA). https://arena.gov.au/about/funding-strategy-investment-plan/renewable-energy-in-residential-commercial-buildings/. Accessed 6 Feb 2018
2. City of Melbourne (2015) Homes for people: housing strategy 2014–2018. City of Melbourne. https://www.melbourne.vic.gov.au/SiteCollectionDocuments/homes-for-people-housing-strategy.pdf. Accessed 18 Feb 2018
3. Cohen R, Bannister P (2017) Why is Australia better than the UK at building energy efficiency? Energy World Jan 2017:18–20
4. DEWHA (2008) Energy use in the Australian residential sector 1986–2020. Australian Government Department of the Environment, Water, Heritage and the Arts (DEWHA), Canberra ACT
5. Rogers R (1997) Cities for a small planet. Faber & Faber, London
6. Williamson T, Radford A, Bennetts H (2003) Understanding sustainable architecture. Spon Press, London

Dr. Trivess Moore, B.Sc. (Env.), Ph.D. is a research fellow with expertise in energy efficiency, renewable energies, the built environment and transitioning to a low carbon future.

Dr. Mary Myla Andamon, B.Sc. Arch., Ph.D. is a lecturer and a building scientist with expertise in indoor and outdoor thermal comfort studies, urban microclimate and building environmental monitoring and performance assessments.

Assoc. Prof. Priyadarsini Rajagopalan, B.Tech. Arch. Eng., M.Bld.Sc., Ph.D. is an associate professor with expertise in building energy efficiency and energy benchmarking, indoor environmental performance and urban microclimate.

MIX
Papier aus verantwortungsvollen Quellen
Paper from responsible sources
FSC® C105338

If you have any concerns about our products,
you can contact us on
ProductSafety@springernature.com

In case Publisher is established outside the EU,
the EU authorized representative is:
Springer Nature Customer Service Center GmbH
Europaplatz 3, 69115 Heidelberg, Germany

Printed by Libri Plureos GmbH
in Hamburg, Germany